RECENT ADVANCES IN COMPRESSED SENSING:
DISCRETE UNCERTAINTY PRINCIPLES AND
FAST HYPERSPECTRAL IMAGING

I. Introduction

Good decisions are typically informed by relevant data. Unfortunately, many settings exhibit a bottleneck at the data-collection step, which presents an issue in the face of urgency. Over the last decade, this limitation has spurred research in speeding up data collection with a developing theory called *compressed sensing*.

The original motivation for compressed sensing research came from applications to medical imaging, e.g., magnetic resonance imaging (MRI). Since the early 1980s, MRI has granted doctors the ability to distinguish between healthy tissue and cancerous tumors without the need for invasive surgical procedures. Unfortunately, MRI machines take a significant amount of scanning time to produce high-resolution images, which are often necessary for diagnostics. The longer a patient is required to remain still, the more likely she is to move, which causes irreparable distortions in the final image. In the last decade, Candès, Romberg and Tao [1] showed that high-resolution images, similar to those produced using MRI, can be reconstructed from very few measurements, which in turn reduces the time required to collect the necessary data. To achieve this speedup, Candès et al. leveraged the fact that the desired image is nearly sparse in the wavelet domain [2].

More generally, compressed sensing seeks to solve the linear system $\Phi x = y$, where $\Phi \in \mathbb{C}^{M \times N}$ with $M \ll N$. Of course, since $M \ll N$, this underdetermined linear system is impossible to uniquely solve using traditional linear algebra. However, if

1

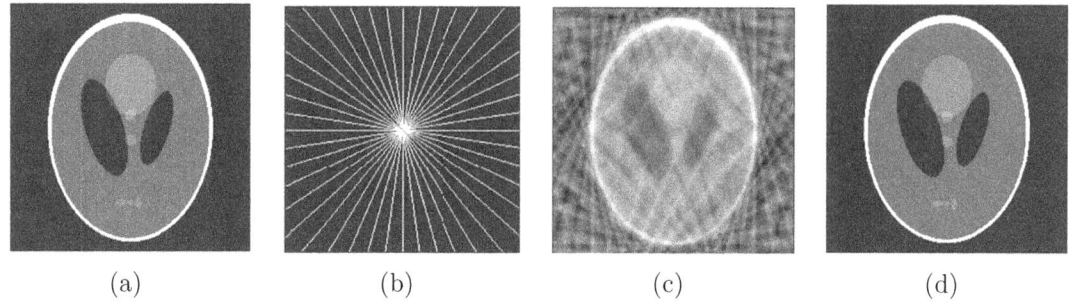

(a)	(b)	(c)	(d)

Figure 1. (a) The Shepp–Logan phantom test image. (b) White pixels denote sample locations of (a) in the Fourier domain. (c) The test image projected onto the span of the sampled Fourier modes. (d) Reconstruction of (a) by minimizing total variation subject to the Fourier samples. Perhaps surprisingly, the reconstruction is exact [1].

we make additional assumptions about x, they might be combined with the linear data to uniquely determine the original vector x. For example, one might assume x is K-sparse, i.e., at most K entries of x are nonzero. In many settings, this is a valid signal model; for example, JPEG2000 exploits the fact that natural images are nearly sparse in the wavelet domain [3, 4].

For a demonstration of compressed sensing, see Figure 1. Here, an image is sampled in only a few locations in the Fourier domain. Traditionally, one might project the image onto the span of these Fourier modes to get Figure 1(c), thereby producing significant artifacts that would prevent proper diagnostics. However, the image can be exactly recovered by instead minimizing a certain convex objective function subject to the Fourier samples. This is the compressed sensing solution that delivers high-resolution imagery from very few measurements. To see why this works, first note that the sensing matrix must be special in order to provide complete information. Indeed, if x is sparse but the rows of Φ happen to be orthogonal to x, then Φx fails to distinguish x from the zero vector. This problem can be averted if Φ satisfies the $(2K, \delta)$-*restricted isometry property* (RIP), that is, if

$$(1 - \delta)\|x\|_2^2 \leq \|\Phi x\|_2^2 \leq (1 + \delta)\|x\|_2^2$$

for every $2K$-sparse x. The main result in compressed sensing is that solving

$$\arg\min \|x\|_1 \qquad \text{subject to} \qquad \Phi x = y$$

is equivalent to finding the sparsest x such that $\Phi x = y$ provided Φ satisfies the $(2K, \delta)$-RIP with $\delta < \sqrt{2} - 1$ and the sparsest such x is K-sparse [5]. Thus, not only do RIP matrices provide complete information about any K-sparse vector, they also allow for reconstruction by ℓ_1 minimization, which can be implemented using linear programming. Furthermore, if Φ is a random matrix (e.g., Φ has iid Gaussian entries), then with high probability, Φ satisfies the $(2K, \delta)$-RIP provided $K = O_\delta(M/\operatorname{polylog} N)$ [6]. Therefore, the number of measurements required to efficiently recover x scales linearly with the signal complexity K and does not significantly depend on the size N of the signal.

The purpose of this thesis is to further develop the theory of compressed sensing and related topics. In particular, we provide a new uncertainty principle (which quantifies the sparsity level of any function relative to its Fourier transform) and we also demonstrate the applicability of compressed sensing theory to fast hyperspectral imaging. The following section details the applicability of these results.

1.1 Applications of uncertainty principles

Uncertainty principles have been studied with increasing frequency for over the last sixty years [7]. The most basic form of the uncertainty principle is the following assertion: *A signal cannot be highly concentrated in both time and frequency,* illustrated in Figure 2. In particular, a Fourier transform pair x and $\hat{x}$ cannot be simultaneously sparse [7, 8, 9]. It wasn't until more recently that uncertainty principles were applied to sparse signal recovery problems, most notably by Donoho and

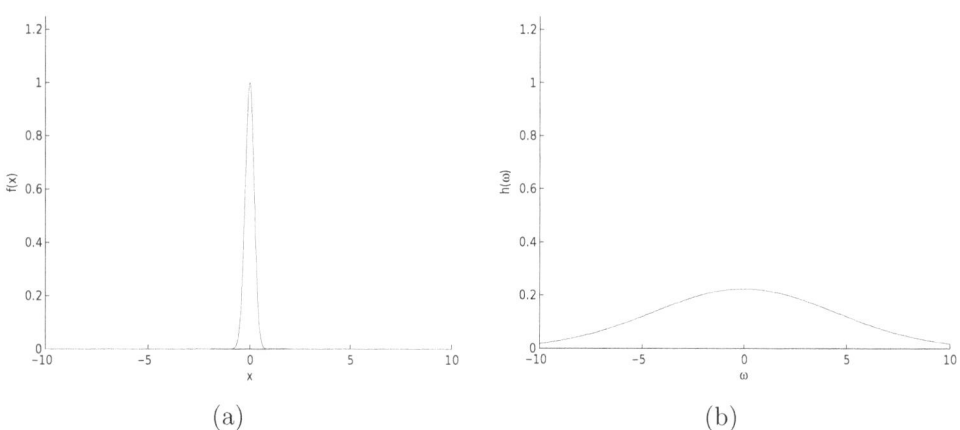

(a)   (b)

Figure 2. (a) **The function** $f(x) = \exp(-10x^2)$. (b) **The Fourier transform of** f, $h(\omega) = \frac{1}{2\sqrt{5}}\exp(-\omega^2/40)$. **Notice that** f **is narrow while** h **is wide. The uncertainty principle contends that no function is simultaneously narrow in both the time and frequency domains.**

Stark in [8].

In this thesis, we develop a new version of the uncertainty principle for discrete functions and discuss its applications to demixing problems and to the fast detection of sparse signals. We briefly describe these problems below:

The demixing problem.

Digital signals are often corrupted by noise when transmitted. Thankfully, it is sometimes possible to distinguish the noise from the desired signal. Assume the desired signal x is sparse in the Fourier domain. If x has been corrupted by noise ϵ that is sparse in the identity basis (think speckle), then we will observe $z = x + \epsilon$. The goal of demixing is to separate x from ϵ. In [10], Tropp showed that demixing can be accomplished by solving

$$[I\ F]v = Fz. \tag{1}$$

4

The desired solution $v^\star$ to (1) is a concatenation of Fx and ϵ, i.e.,

$$v^\star = \begin{bmatrix} Fx \\ \epsilon \end{bmatrix}.$$

Since $v^\star$ is sparse, this problem may be solved using ideas from compressed sensing. In particular, $v^\star$ is likely the 1-norm minimizer subject to (1). However, suppose there is a signal f that is sparse in both the identity and Fourier bases. Then $F(x+f)$ and $\epsilon - f$ are both sparse. Therefore,

$$v = \begin{bmatrix} F(x+f) \\ \epsilon - f \end{bmatrix}$$

is also a sparse solution to (1). As such, we cannot guarantee recovery of $v^\star$ using compressed sensing methods because $v^\star$ is not the only sparse solution.

Indeed, we need to understand how sparse a function can be in both the identity and Fourier bases in order to establish the inherent limitations on the demixing problem. This is one motivation behind our study of uncertainty principles.

Fast detection of sparse signals.

Signal detection is a well-studied problem with a variety of applications. The goal of signal detection is to distinguish between an information-bearing signal and mere noise, making it a critical component of signals intelligence. Ideally, detection should be performed as quickly as possible so as to promptly initialize further processing and enable a timely decision. Hence, it is desirable to use as little computation as possible to detect an information-bearing signal. Assume that a signal x is K-sparse. Then by the uncertainty principle, Fx is not highly sparse, and so only a few random samples in the Fourier domain should suffice to detect the signal. In Chapter III, we will show that only $O(K)$ random samples are required to detect a sparse signal, and our detection method requires only $O(K \log N)$ time.

1.2 Applications to fast hyperspectral imaging

A standard digital camera captures and stores images using only red, green and blue light channels. Combinations of these frequencies of light are then used by computer screens to display photographs for human consumption. Hyperspectral imaging involves collecting data at multiple wavelengths rather than just these three. The advantage of collecting multiple spectral bands is the ability to glean information about the chemical composition of the observed object [11]. As a defense application, hyperspectral imaging is particularly important for analyzing the composition of detonations in war zones [12]. With this application in mind, we seek to observe brief and localized events. The fact that the event is brief is a disadvantage since conventional hyperspectral imaging platforms might not have time to capture the desired image. On the other hand, the advantage of localized events is that they are spatially sparse, allowing for compressed sensing to possibly overcome the event's brevity.

Assume $X = f(x, y, \lambda)$ is the data cube we would like to observe, where x and y are the spatial coordinates and λ is a spectral coordinate, often referred to as a spectral band. Conventional remote sensing platforms do not collect hyperspectral data without some sort of scanning method. The three main scanning methods are spectral, point, and line scanning [13]. Spectral scanning involves the platform only obtaining information for a particular spectral band at a time. Point scanning (or whisk-broom) methods collect all light channels at one point in space at a time. Line scanning (or push-broom) methods observe an entire line in space at a time, moving across space. Therefore, this method collects two dimensional images with each measurement, one spatial dimension and a spectral dimension [13]. The disadvantage

6

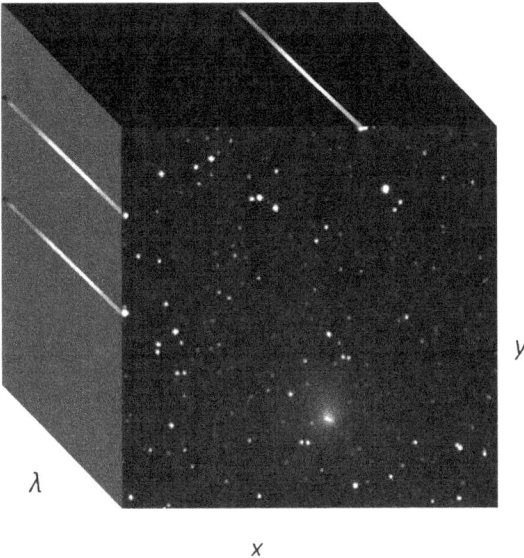

Figure 3. Example of data cube of hyperspectral imagery of stars. The spatial location of a star is represented with x- and y-coordinates. The λ axis represents the spectral bands of the data cube. The data cube is sparse since the distribution of stars in the night sky is sparse and spectral bands are zero throughout empty space. Image of stars from [14].

with each of these scanning methods is that they require a large amount of time to capture the data cube X [11].

Suppose X is a hyperspectral image of the night sky as shown in Figure 3. We know that stars are sparsely distributed throughout space, and so we may assume that X is sparse. As such, compressed sensing methods might allow us to bypass the limitations of conventional scanning. Along these lines, an alternative hyperspectral imaging model, proposed by [15], uses a micro-mirror array (MMA) to capture the entire data cube instantaneously (see Figure 4). Specifically, the MMA either reflects all light at a point in space through a prism onto a charge-coupled device (CCD) or reflects the light away from the CCD.

Given compressive measurements of the data cube, we need to apply a reconstruction algorithm. To this end, we will choose an algorithm that exploits the additional structure of the desired data cube. Indeed, notice in Figure 3 that the nonzero en-

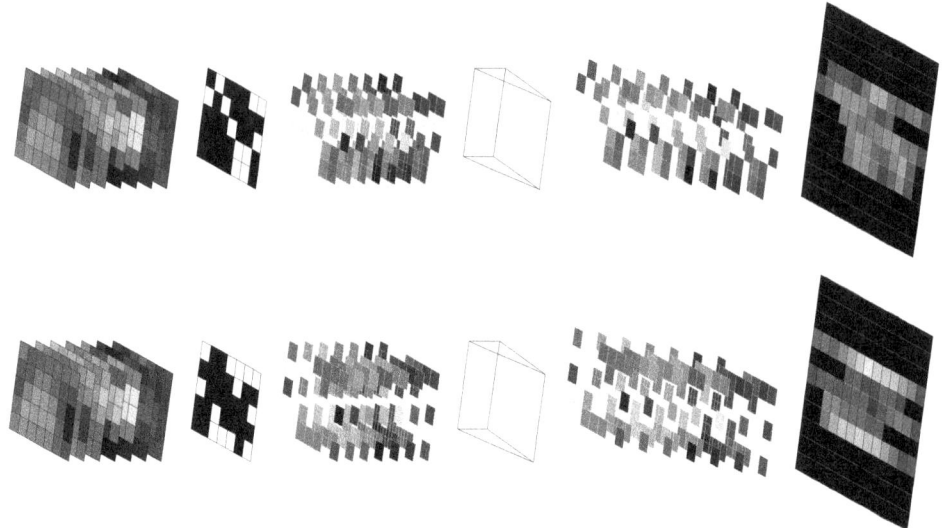

Figure 4. Example of a compressive remote sensing platform. Each exposure uses a different configuration of the micro-mirror array (MMA). A mirror in the MMA either reflects all light towards the prism or reflects all light away from the prism. After the light is dispersed by the prism, it is collected on the charge-coupled device.

tries of the data cube are in batches with common spatial coordinates. We call these batches, *blocks* and we say the data cube exhibits *block sparsity*. Block sparsity is more informative than regular sparsity, and recovery algorithms that use this information tend to perform better [16, 17]. One such recovery algorithm, developed in [16], is *block orthogonal matching pursuit* (BOMP). As the name suggests, this algorithm is a block version of a sparsity-based reconstruction algorithm called *orthogonal matching pursuit* (OMP). In Chapter IV, we apply this algorithm to simulated data cubes of star data to demonstrate the plausibility of data cube reconstruction from few hyperspectral exposures. In particular, our simulations suggest that one can reconstruct data cubes like the one depicted in Figure 3 with less than a third of the exposures required by conventional scanning methods.

1.3 Outline

This introduction serves to build the context of the theory developed in Chapter II, as well as the real-world problems that will benefit from our results. In Chapter II, we will discuss the 0-norm uncertainty principle and give a concise proof of the Donoho–Stark uncertainty principle developed in [8]. Additionally, we introduce and characterize equality in a new mixed-norm uncertainty principle. Lastly, we derive a Fourier transform pair that achieves near equality in the Donoho–Stark uncertainty principle as well as the mixed-norm uncertainty principle.

In Chapter III, we consider two applications of the mixed-norm uncertainty principle. First, we show that demixing problems cannot break the square-root bottleneck, i.e., in the worst case one can only stably demix K-sparse signals if $K = O(\sqrt{N})$, where N is the signal dimension [10, 18]. Second, we show how to detect K-sparse signals from only $O(K)$ measurements.

In Chapter IV, we focus on the hyperspectral imaging problem. First we use combinatorial designs to devise a sequence of micro-mirror orientations that allow us to reconstruct the desired hyperspectral imagery with block orthogonal matching pursuit. Second, we provide simulation results that demonstrate the feasibility of compressive hyperspectral imaging.

We conclude in Chapter V with ideas for future work.

II. Discrete uncertainty principles

In this chapter, we discuss three discrete uncertainty principles. We begin with a proof of the 0-norm uncertainty principle [19] and a new efficient proof of the Donoho–Stark uncertainty principle (Theorem 2 in [8]). Next, we discuss a new uncertainty principle that we call the *mixed-norm uncertainty principle*. Later we prove a surprising result that equality in the mixed-norm uncertainty principle is achieved precisely when equality is achieved in the 0-norm uncertainty principle. Lastly, we show that a discretized and periodized Gaussian achieves near equality in both the Donoho–Stark uncertainty principle and the mixed-norm uncertainty principle.

2.1 Background

Before we begin, we will introduce some notation and definitions. Let G be an abelian group and let $\mathbb{T}$ denote the complex unit circle. Define a character of G to be a function $\chi : G \to \mathbb{T}$ such that

$$\chi(g_1 + g_2) = \chi(g_1)\chi(g_2) \qquad \forall g_1, g_2 \in G. \tag{2}$$

Let $\widehat{G}$ be the set of all homomorphisms χ that satisfy (2). The dual of G, namely $\widehat{G}$, is also an abelian group (see Appendix 1.1). Let $\ell(G)$ be the Hilbert space of all functions from G to $\mathbb{C}$. We denote the Fourier transform of a function $f \in \ell(G)$ by $\hat{f} \in \ell(\widehat{G})$. We formally define the Fourier transform from $\ell(G)$ to $\ell(\widehat{G})$ below.

Definition 1. *The Fourier transform $F \colon \ell(G) \to \ell(\widehat{G})$ and its inverse are given by*

$$(Ff)[\chi] := \frac{1}{\sqrt{|G|}} \sum_{g \in G} f[g]\overline{\chi[g]}, \qquad (F^{-1}h)[g] := \frac{1}{\sqrt{|G|}} \sum_{\chi \in \widehat{G}} h[\chi]\chi[g],$$

respectively.

Another useful definition is the support of a function, denoted $\mathrm{supp}(\cdot)$. The support of a function f on its domain D is defined as

$$\mathrm{supp}(f) := \{x \in D : f(x) \neq 0\}.$$

The 0-norm is then given by $\|f\|_0 = |\mathrm{supp}(f)|$. We note that the 0-norm is not technically a norm, since it violates homogeneity.

2.2 The 0-norm uncertainty principle

In this section, we introduce the 0-norm uncertainty principle (see [19]). Throughout this thesis, we will assume that $|G| = N$.

Theorem 1. *Given an abelian group G of size N and $f \in \ell(G)$, we have*

$$|\mathrm{supp}(f)||\mathrm{supp}(\hat{f})| \geq N.$$

Proof. By the triangle inequality and Definition 1,

$$\max_{\chi \in \widehat{G}} |\hat{f}[\chi]| \leq \left| \frac{1}{\sqrt{N}} \sum_{g \in G} f[g] \right| \leq \frac{1}{\sqrt{N}} \sum_{g \in G} |f[g]|. \tag{3}$$

For any $y \in \mathbb{C}$, define $\mathrm{sgn}(y)$ as

$$\mathrm{sgn}(y) := \begin{cases} y/|y| & \text{if } y \neq 0 \\ 0 & \text{if } y = 0. \end{cases}$$

We know that $|f[g]|^2 = f[g]\overline{f[g]}$. Dividing by $|f[g]|$ yields

$$|f[g]| = f[g]\frac{\overline{f[g]}}{|f[g]|} = f[g]\overline{\operatorname{sgn} f[g]}. \tag{4}$$

Substituting (4) into (3) gives

$$\max_{\chi \in \widehat{G}} \left|\hat{f}[\chi]\right| \leq \frac{1}{\sqrt{N}} \sum_{g \in G} f[g]\overline{\operatorname{sgn} f[g]} = \frac{1}{\sqrt{N}} \langle f, \operatorname{sgn} f \rangle \leq \frac{1}{\sqrt{N}} \|f\|_2 \|\operatorname{sgn} f\|_2, \tag{5}$$

where the last inequality is true by the Cauchy–Schwarz inequality. By definition of $\|\cdot\|_2$, we have

$$\|f\|_2 \|\operatorname{sgn} f\|_2 = \|f\|_2 \left(\sum_{g \in G} |\operatorname{sgn} f[g]|^2\right)^{\frac{1}{2}} = \|f\|_2 |\operatorname{supp}(f)|^{\frac{1}{2}} = \|\hat{f}\|_2 |\operatorname{supp}(f)|^{\frac{1}{2}}, \tag{6}$$

where the last equality is a result of Plancherel's theroem. Note that for all $y \in \ell(G)$, $\|y\|_2 \leq |\operatorname{supp}(y)|^{1/2}\|y\|_\infty$. Therefore,

$$\|\hat{f}\|_2 \leq |\operatorname{supp}(\hat{f})|^{1/2} \max_{\chi \in \widehat{G}} |\hat{f}[\chi]|. \tag{7}$$

Combining (5), (6), and (7) yields

$$\max_{\chi \in \widehat{G}} |\hat{f}[\chi]| \leq \frac{1}{\sqrt{N}} \max_{\chi \in \widehat{G}} |\hat{f}[\chi]|| \operatorname{supp}(f)|^{1/2}| \operatorname{supp}(\hat{f})|^{1/2}.$$

Taking the square and rearranging then gives

$$|\operatorname{supp}(f)|| \operatorname{supp}(\hat{f})| \geq N. \qquad \square$$

We will study two robust versions of Theorem 1. First is the uncertainty principle introduced by Donoho and Stark [8]. The second theorem is in terms of a numerically robust analog of the 0-norm called numerical sparsity (see Definition 3).

2.3   The Donoho–Stark uncertainty principle

In this section, we introduce the Donoho–Stark uncertainty principle and provide a more efficient proof than the one provided in [8]. Define $T \subseteq G$, $W \subseteq \widehat{G}$ and let $P_T, P_W : \ell(G) \to \ell(G)$ be time- and frequency-limiting operators defined by

$$(P_T f)[t] := \begin{cases} f[t] & \text{if } t \in T \\ 0 & \text{otherwise} \end{cases}$$

and

$$(P_W f)[t] := \frac{1}{\sqrt{N}} \sum_{w \in W} e^{\frac{2\pi i s w}{N}} \hat{f}[w]$$

respectively. The projection operator P_T removes any part of the function f that is supported outside the index set T. Similarly, P_W filters f so that $\hat{f}$ is supported on the index set W.

Theorem 2 (Donoho–Stark [8]). *Let G be an abelian group of size N and suppose $f \in \ell(G)$ is concentrated in both time and frequency:*

$$\|f - P_T f\|_2 \le \epsilon_T \|f\|_2, \qquad \|f - P_W f\|_2 \le \epsilon_W \|f\|_2,$$

for some $\epsilon_T, \epsilon_W \ge 0$. Then

$$|T||W| \ge N\big(1 - (\epsilon_T + \epsilon_W)\big)^2.$$

Proof. Assume without loss of generality that $\|f\| = 1$. By the triangle inequality,

$$1 - \|P_W P_T f\| \le \|f - P_W P_T f\| \le \|f - P_W f\| + \|P_W(f - P_T f)\|. \tag{8}$$

13

Since P_W is a projector,

$$\|f - P_W f\| + \|P_W(f - P_T f)\| \leq \|f - P_W f\| + \|f - P_T f\| \leq \epsilon_W + \epsilon_T. \qquad (9)$$

Combining (8) and (9), and then rearranging produces

$$1 - (\epsilon_W + \epsilon_T) \leq \|P_W P_T f\| \leq \|P_W P_T\|_{2 \to 2} \leq \|P_W P_T\|_F,$$

where the second inequality follows from the definition of the induced norm and $\|\cdot\|_F$ denotes the Frobenius norm. We claim $\|P_W P_T\|_F^2 = |W||T|/N$, which implies the result.

To prove our claim, define $D_S := \mathrm{diag}(\mathbf{1}_S)$, where $\mathbf{1}_S$ denotes the indicator function of the set S. Note that $P_T = D_T$ and $P_W = F^{-1} D_W F$. Therefore,

$$\|P_W P_T\|_F^2 = \|F^{-1} D_W F D_T\|_F^2 = \|D_W F D_T\|_F^2 = \sum_{\substack{i \in W \\ j \in T}} |F[i,j]|^2 = \frac{|T||W|}{N},$$

completing the proof. $\qquad\qquad\square$

2.4 Near equality in the Donoho–Stark uncertainty principle

In this section, we show that a discretized and periodized Gaussian function achieves near equality in the Donoho–Stark uncertainty principle. In particular, we have the following result:

Theorem 3. *For every $\epsilon > 0$, there exists $C > 0$ such that for every sufficiently large N, there exists $f \in \ell(\mathbb{Z}_N)$, $T \subset \mathbb{Z}_N$, $W \subset \widehat{\mathbb{Z}_N}$ such that*

$$\|f - P_T f\|_2 \leq \epsilon\|f\|_2, \qquad \|f - P_W f\|_2 \leq \epsilon\|f\|_2$$

14

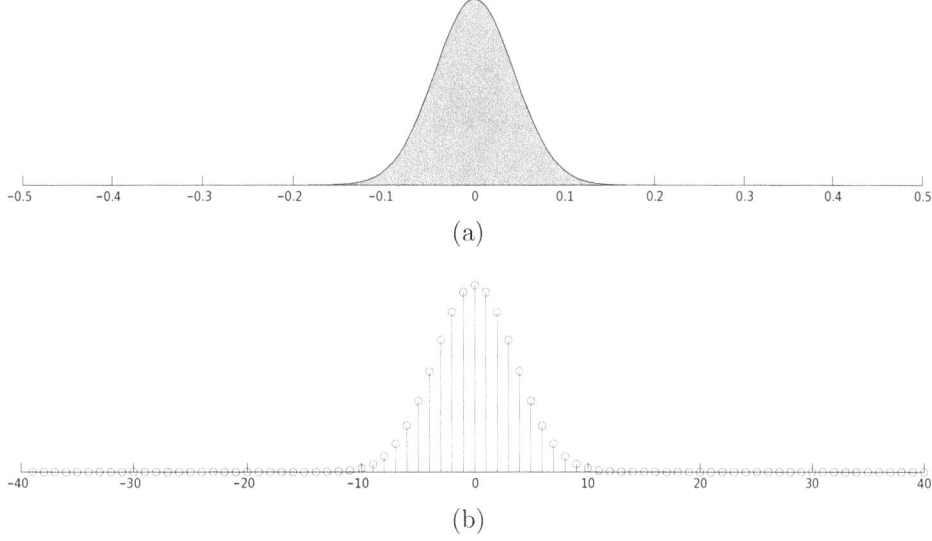

(a)

(b)

Figure 5. (a) An example of a Gaussian function $f(x) = e^{-N\pi x^2}$, with $N = 81$. (b) The Gaussian function from (a) sampled at every multiple of $1/N$ and periodized. This function is an eigenvector of the discrete Fourier transform, and it has around $\sqrt{N}$ large entries. We leverage these facts to demonstrate near equality in both the Donoho–Stark and mixed-norm uncertainty principles.

and

$$|T||W| \leq CN \log N. \tag{10}$$

Such a construction will demonstrate that Theorem 2 is nearly tight for all sufficiently large N (see Figure 5). Before providing the construction rigorously, we begin with a definition:

Definition 2. *Let Schwarz space be the set*

$$\mathcal{S} = \left\{ f \in C^\infty(\mathbb{R}) : \sup_{x \in \mathbb{R}} |x^\alpha f^{(\beta)}(x)| < \infty \ \forall \alpha, \beta \right\}.$$

Theorem 4. *For all $f \in \mathcal{S}$, if*

$$h[j] = \sum_{j' \in \mathbb{Z}} f\left(\frac{j}{N} + j'\right),$$

15

then

$$(F_N h)[k] = \sqrt{N} \sum_{k' \in \mathbb{Z}} \hat{f}\left(k + k'N\right),$$

where F_N denotes the Fourier transform on $\mathbb{Z}_N$.

Proof. First, by the definition of F_N

$$(F_N h)[k] = \frac{1}{\sqrt{N}} \sum_{j \in \mathbb{Z}_N} \left(\sum_{j' \in \mathbb{Z}} f\left(\frac{j}{N} + j'\right)\right) e^{\frac{-2\pi i j k}{N}}.$$

Apply Corollary 1 (Appendix 1.2) and rearrange to get

$$(F_N h)[k] = \frac{1}{\sqrt{N}} \sum_{j \in \mathbb{Z}_N} \left(\sum_{q \in \mathbb{Z}} (F_{\mathbb{R}} f)[q] e^{2\pi i j q/N}\right) e^{-2\pi i j k/N}$$

$$= \frac{1}{\sqrt{N}} \sum_{q \in \mathbb{Z}} (F_{\mathbb{R}} f)[q] \sum_{j \in \mathbb{Z}_N} \left(e^{2\pi i (q-k)/N}\right)^j.$$

Then the geometric sum formula (see Claim 9 in Appendix 1.4) implies

$$\frac{1}{\sqrt{N}} \sum_{q \in \mathbb{Z}} (F_{\mathbb{R}} f)[q] \sum_{j \in \mathbb{Z}_N} \left(e^{2\pi i (q-k)/N}\right)^j = \sqrt{N} \sum_{q = k + N\mathbb{Z}} (F_{\mathbb{R}} f)[q]$$

Let $k' := (q - k)/N$. By change of variables, we are done:

$$(F_N H)[k] = \sqrt{N} \sum_{k' \in \mathbb{Z}} (F_{\mathbb{R}} f)\left(k + k'N\right). \qquad \square$$

Armed with Theorem 4 and the background information in Appendix 1.3, we can find the Fourier transform pair we seek:

Theorem 5. *If $f_s(x) := \frac{2^{\frac{1}{4}}}{\sqrt{s}} e^{-\pi\left(\frac{x}{s}\right)^2}$, then $(Ff)_s(\xi) := f_{\frac{1}{s}}(\xi)$.*

Proof. By the definition of f_s, we know that for some function h,

$$f_s(\xi) = F\left(h\left(\frac{x}{s}\right)\right) = s(Fh)(s\xi)$$

16

where the last equality is due to Claim 8. By definition of h, we get

$$s(Fh)(s\xi) = 2^{\frac{1}{4}}\sqrt{s}e^{-\pi(\xi s)^2} = f_{1/s}(\xi). \qquad \square$$

Given the Fourier transform pair f_s and $f_{1/s}$ and Theorem 4, we can derive the following Fourier transform pair.

Lemma 1. *Denote*

$$h_s[n] := \sum_{n'\in\mathbb{Z}} f_s\left(\frac{n}{N} + n'\right).$$

Then, $Fh_s[n] = h_{\frac{1}{Ns}}[n]$.

Proof. Given f_s and the definition of h_s,

$$h_s[n] := \sum_{n'\in\mathbb{Z}} f_s\left(\frac{n}{N} + n'\right) = \sum_{n'\in\mathbb{Z}} \frac{2^{\frac{1}{4}}}{\sqrt{s}} e^{-\pi\left(\frac{n}{N}+n'\right)^2\left(\frac{1}{s}\right)^2}.$$

Theorem 4 and Lemma 5 then give

$$Fh_s[n] := \sqrt{N} \sum_{n'\in\mathbb{Z}} \widehat{f_s}(n + n'N) = \sqrt{N} \sum_{n'\in\mathbb{Z}} f_{1/s}(n + n'N).$$

By definition of $f_{1/s}$

$$\sqrt{N} \sum_{n'\in\mathbb{Z}} f_{1/s}(n + n'N) = \sqrt{N} \sum_{n'\in\mathbb{Z}} 2^{\frac{1}{4}}\sqrt{s}e^{-\pi(S(n+Nn'))^2}$$

$$= 2^{\frac{1}{4}}\sqrt{Ns} \sum_{n'\in\mathbb{Z}} e^{-\pi\left(Ns\left(\frac{n}{N}+n'\right)\right)^2} = h_{\frac{1}{Ns}}[n] \qquad \square$$

Let $h^{(K)}$ be defined as follows:

$$h^{(K)}[n] := \sqrt{\frac{N}{K}} \sum_{n'\in\mathbb{Z}} e^{-\pi\left(\frac{n+n'N}{K}\right)^2}. \tag{11}$$

17

By Lemma 1, we can see that that the Fourier transform of $h^{(K)}$ is simply $h^{(N/K)}$. The following facts about $h^{(K)}$ will be useful for proving near equality.

Lemma 2. *Define $h^{(K)}$ as in (11). Then*

$$\|h^{(K)}\|_2^2 \geq \frac{N}{\sqrt{2}} - \frac{N}{K}.$$

Proof. First, we expand $|w|^2 = w\overline{w}$ to get

$$\|h^{(K)}\|_2^2 = \sum_{n \in \mathbb{Z}_N} \left| \sqrt{\frac{N}{K}} \sum_{n' \in \mathbb{Z}} e^{-\pi \left(\frac{n+n'N}{K}\right)^2} \right|^2 = \frac{N}{K} \sum_{n \in G} \sum_{n' \in \mathbb{Z}} \sum_{n'' \in \mathbb{Z}} e^{-\pi \left[\left(\frac{n+n'N}{K}\right)^2 + \left(\frac{n+n''N}{K}\right)^2 \right]}.$$

Since all of the terms in the sum are nonnegative, we may infer a lower bound by discarding the terms for which $n'' \neq n'$. This yields the following:

$$\|h^{(K)}\|_2^2 \geq \frac{N}{K} \sum_{n \in G} \sum_{n' \in \mathbb{Z}} e^{-2\pi \left(\frac{n+n'N}{K}\right)^2} = \frac{N}{K} \sum_{m \in \mathbb{Z}} e^{-2\pi \frac{m^2}{K^2}} \geq \frac{N}{K} \left(\int_{-\infty}^{\infty} e^{-2\pi \frac{x^2}{K^2}} dx - 1 \right),$$

where the last inequality follows from an integral comparison. The result then follows from computing the integral. $\qquad \square$

Lemma 3. *If $h^{(K)}$ is defined as in (11) and $0 \leq n \leq n' \leq \frac{N}{2}$, then $h^{(K)}[n] \geq h^{(K)}[n']$.*

Proof. It suffices to show that $e^{-\pi \left(\frac{n}{K}\right)^2} \geq e^{-\pi \left(\frac{n'}{K}\right)^2} + e^{-\pi \left(\frac{n'-N}{K}\right)^2}$, which is true if $e^{-\pi \left(\frac{n}{K}\right)^2} \geq e^{-\pi \left(\frac{n+1}{K}\right)^2} + e^{-\pi \left(\frac{N/2}{K}\right)^2}$. Define $a := e^{-\frac{\pi}{K^2}}$. Then it suffices to show that

$$a^{n^2} \geq a^{(n+1)^2} + a^{\left(\frac{N}{2}\right)^2}.$$

Dividing both sides by a^{n^2} yields

$$1 \geq a^{2n+1} + a^{\left(\frac{N}{2}\right)-n^2} = a^{2n+1} + a^{\left(\frac{N}{2}-n\right)\left(\frac{N}{2}+n\right)} \geq a + a^{\frac{N}{2}}.$$

The last inequality is from the fact that $2n + 1 \geq 1$ and $\left(\frac{N}{2} - n\right)\left(\frac{N}{2} + n\right) \geq \frac{N}{2}$. The above implies that $e^{-\frac{\pi}{K}^2} = a \geq \frac{1}{2}$, which is true because of our assumption that $K \geq 4$. $\qquad\qquad\square$

We are now ready to prove the main result of this section.

Proof of Theorem 3. Take $f = h^{(K)}$ with $K = \sqrt{N}$, and so $\hat{f} = f$. It is then reasonable to take $T = W$ and $\epsilon_T = \epsilon_W$. With this, in order to show that Theorem 2 is tight for $h^{(K)}$, it suffices to find T such that (10) is satisfied and $\|h^{(K)} - P_T h^{(K)}\| < \epsilon_T \|h^{(K)}\|$.

We will find a $p > 0$ such that

$$f[p] < \delta \tag{12}$$

for some δ. Then $T = \{n \in \mathbb{Z}_N : |n| < p\}$. By Lemma 3,

$$f[n] \leq f[p] < \delta \qquad \forall n \notin T$$

and so

$$\|f - P_T f\|^2 \leq (N - |T|)\delta^2 \leq N\delta^2 \leq 2\delta^2 \|f\|^2,$$

where the last inequality follows from Lemma 2 and the assumption that $N \geq 16$. Thus, $\|f - P_T f\| \leq \epsilon_T \|f\|$ for $\epsilon_T = \sqrt{2}\delta$.

Also, by our choices of W, f, and ϵ_W we have

$$\|f - P_W f\| = \|f - F^{-1} P_T F f\| = \|F^{-1}(Ff - P_T F f)\| = \|Ff - P_T F f\|.$$

We know that $f = Ff$. Therefore,

$$\|Ff - P_T F f\| = \|f - P_T f\| \leq \epsilon_T \|f\| = \epsilon_W \|f\|.$$

19

As such, f will satisfy the hypotheses of Theorem 2. If, in addition, we find a small p satisfying (12), then we will have $|T||W| = (2p-1)^2$ small as well, so Theorem 2 is nearly tight for this choice of f.

By Lemma 3 and symmetry, $h^{(K)}[p] \geq h^{(K)}[n] \ \forall n \in \{p, \ldots, \ N - p\}$. Thus, it suffices to show a smallest p possible such that $h^{(K)}[p] < \delta$ and $p = O(K)$. In order to find p, we will find an upper bound on $h^{(K)}$ in terms of n and then set $n = p$.

$$h^{(K)}[n] = \sqrt{\frac{N}{K}} \left(\sum_{\substack{n' \in \mathbb{Z} \\ \left|\frac{n+n'N}{K}\right| \geq t}} e^{-\pi\left(\frac{n+n'N}{K}\right)^2} + \sum_{\substack{n' \in \mathbb{Z} \\ \left|\frac{n+n'N}{K}\right| < t}} e^{-\pi\left(\frac{n+n'N}{K}\right)^2} \right) \tag{13}$$

Assume $t \geq \sqrt{N}$. We will bound each term in (13) separately, beginning with the first term. Let $n = \alpha\sqrt{N}$, where $\alpha \leq \frac{1}{2}\sqrt{N}$. Then,

$$\sum_{\substack{n' \in \mathbb{Z} \\ \left|\frac{n+n'N}{K}\right| \geq t}} e^{-\pi\left(\frac{n+n'N}{K}\right)^2} = \sum_{\substack{n' \in \mathbb{Z} \\ \left|\alpha+n'\sqrt{N}\right| \geq t}} e^{-\pi(\alpha+n'\sqrt{N})^2} = \frac{1}{\sqrt{N}} \sum_{\substack{n' \in \mathbb{Z} \\ \left|\alpha+n'\sqrt{N}\right| \geq t}} \sqrt{N} e^{-\pi(\alpha+n'\sqrt{N})^2}.$$

By our assumption that $t \geq \sqrt{N}$, the above is bounded by

$$\frac{1}{\sqrt{N}} \int_{-\infty}^{\infty} e^{-\pi y^2} dy = \frac{1}{\sqrt{N}}.$$

Now, we will bound the second term in (13). Because $\alpha \leq \frac{1}{2}\sqrt{N}$, $e^{-\pi(\alpha n'\sqrt{N})^2}$ is maximized for $n' = 0$. Assume, $t = \alpha\sqrt{N}$. Then,

$$\left|\frac{n+n'N}{K}\right| = |\alpha + n'\sqrt{N}| < t = \alpha\sqrt{N}.$$

Because $\alpha > 0$, the above implies $|1/\sqrt{N} + n'/\alpha| < 1$. Therefore

$$-1 < \frac{1}{\sqrt{N}} + \frac{n'}{\alpha} < 1.$$

Solving the inequality for n', we get

$$-\alpha\left(1 + \frac{1}{\sqrt{N}}\right) < n' < -\alpha\left(1 + -\frac{1}{\sqrt{N}}\right).$$

Therefore, $|n'| \leq (1 + o(1))\alpha$. Hence, the second term in (13) is bounded by

$$\#\{n' : |\alpha + n'\sqrt{N}| < t\} \cdot e^{i\pi\alpha^2} \leq 3\alpha e^{-\pi\alpha^2}.$$

Thus, the upper bound for (13) is:

$$h^{(K)}[n] \leq \sqrt{\frac{N}{K}}\left(\frac{1}{\sqrt{N}} + 3\alpha e^{-\pi\alpha^2}\right) = N^{-1/4} + N^{1/4}3\alpha e^{-\pi\alpha^2}. \tag{14}$$

Assuming $N^{-1/4} < \delta/2$, we will find a lower bound on α such that the above is less than δ. Therefore, we need to find an α such that $N^{1/4}3\alpha e^{-\pi\alpha^2} \leq \delta/2$. We know that $\alpha e^{-\pi\alpha^2} < e^{\alpha^2}e^{-\pi\alpha^2}$. Therefore, it suffices to find an α such that

$$(1 - \pi)\alpha^2 < \log\left(\frac{\delta}{6N^{1/4}}\right).$$

Thus, we need α such that

$$\alpha > \left[\frac{1}{\pi - 1}\log\left(\frac{6N^{1/4}}{\delta}\right)\right]^{1/2}.$$

Therefore, for any fixed δ, we have $h^{(K)}[C(\delta)N^{1/2}\log^{1/2}N] < \delta$ for all sufficiently large N. Thus,

$$|T||W| \leq CN \log N$$

as desired. □

2.5 A mixed-norm uncertainty principle

In this section, we define numerical sparsity and use it to formulate a new mixed-norm uncertainty principle. We also provide three different proofs of this uncertainty principle.

Definition 3. *We define the numerical sparsity of $x \in \ell(G)$, denoted $\mathrm{ns}(x)$, as*

$$\mathrm{ns}(x) := \frac{\|x\|_1^2}{\|x\|_2^2}.$$

The concept of numerical sparsity is introduced in [20] as a stable lower bound on the sparsity of an unknown signal, which we prove below. This property of numerical sparsity is useful in the context of our discussion of compressed sensing.

Lemma 4. *Let $f \in \ell(G)$. Then $\mathrm{ns}(f) \leq \|f\|_0$.*

Proof. By the Cauchy–Schwarz inequality we know that

$$\|f\|_1 = \langle \mathrm{sgn}(f), f \rangle \leq \|\mathrm{sgn}(f)\|_2 \|f\|_2 = \sqrt{\|f\|_0} \|f\|_2.$$

Rearranging and Definition 3 then give

$$\mathrm{ns}(f) = \frac{\|f\|_1^2}{\|f\|_2^2} \leq \frac{\|f\|_0 \|f\|_2^2}{\|f\|_2^2} = \|f\|_0,$$

thus, proving the statement. □

Using Definition 3, we have the following uncertainty principle.

Theorem 6 (Mixed-norm uncertainty principle). *Given an abelian group G of size N and $f \in \ell(G)$, we have*

$$\mathrm{ns}(f)\,\mathrm{ns}(\hat{f}) \geq N. \qquad (15)$$

Notice that by Lemma 4, the 0-norm uncertainty principle is immediately implied by the mixed-norm uncertainty principle. We have two similar proofs of Theorem 6 (both of which will be used later to characterize equality in the uncertainty principle) and a third very different proof that uses interesting techniques. We begin with the two similar proofs.

Proof 1 of Theorem 6. Without loss of generality, assume $\|f\|_2 = 1$. By Holder's inequality,

$$\mathrm{ns}(f)\,\mathrm{ns}(\hat{f}) = \frac{\|f\|_1^2}{\|f\|_2^2} \cdot \frac{\|\hat{f}\|_1^2}{\|\hat{f}\|_2^2} \geq \frac{\|f\|_1^2}{\|f\|_2^2} \cdot \frac{\|\hat{f}\|_2^2}{\|\hat{f}\|_\infty^2} = \frac{\|f\|_1^2}{\|\hat{f}\|_\infty^2},$$

where the last equality is due to Plancherel's theorem.

Additionally, we know that

$$\frac{\|\hat{f}\|_\infty}{\|f\|_1} \leq \sup_{f \neq 0} \frac{\|Ff\|_\infty}{\|f\|_1} = \|F\|_{1 \to \infty} = \frac{1}{\sqrt{N}},$$

where F is the Fourier transform operator and $\|\cdot\|_{1 \to \infty}$ is the induced norm.

Therefore

$$\frac{\|f\|_1^2}{\|\hat{f}\|_\infty^2} \geq \frac{1}{\|F\|_{1 \to \infty}^2} = N,$$

and so

$$\mathrm{ns}(f)\,\mathrm{ns}(\hat{f}) \geq N. \qquad \square$$

Proof 2 of Theorem 6. Without loss of generality, assume $\|f\|_2 = 1$. By Hölder's inequality,

$$\mathrm{ns}(f)\,\mathrm{ns}(\hat{f}) = \frac{\|f\|_1^2}{\|f\|_2^2} \cdot \frac{\|\hat{f}\|_1^2}{\|\hat{f}\|_2^2} \geq \frac{\|f\|_2^2}{\|f\|_\infty^2} \cdot \frac{\|\hat{f}\|_1^2}{\|\hat{f}\|_2^2} = \frac{\|\hat{f}\|_1^2}{\|f\|_\infty^2},$$

where the last equality is due to Plancherel's theorem.

Define $y := \hat{f}$, then

$$\frac{\|F^{-1}y\|_\infty}{\|y\|_1} \leq \sup_{y \neq 0} \frac{\|F^{-1}y\|_\infty}{\|y\|_1} = \|F^{-1}\|_{1 \to \infty} = \frac{1}{\sqrt{N}}$$

where F^{-1} is the inverse Fourier transform operator, and so

$$\frac{\|\hat{f}\|_1^2}{\|f\|_\infty^2} = \frac{\|y\|_1^2}{\|F^{-1}y\|_\infty^2} \geq \frac{1}{\|F^{-1}\|_{1 \to \infty}^2} = N.$$

Therefore

$$\mathrm{ns}(f)\,\mathrm{ns}(\hat{f}) \geq N. \qquad \square$$

A third proof of Theorem 6 uses interesting techniques. We start by proving a few lemmas.

Lemma 5. *Let G be a finite abelian group and suppose $f \in \ell(G)$ satisfies $\|f\|_1 \leq C\sqrt{K}\|f\|_2$. Then there exists $T \subseteq G$ of size K such that $\|f - P_T f\|_2 \leq C\|f\|_2$.*

Proof. Let $T_0 \subseteq G$ denote the indices of the K largest entries of f in absolute value, and for each $j \geq 1$, let T_j denote the indices of the K largest entries of f not covered by T_i for $i < j$. Then

$$\|P_{T_{j+1}}f\|_2 \leq \sqrt{K} \max_{g \in T_{j+1}} |f[g]| \leq \sqrt{K} \min_{g \in T_j} |f[g]| \leq \frac{1}{\sqrt{K}}\|P_{T_j}f\|_1$$

for every $j \geq 0$. As such,

$$\|f - P_{T_0}f\|_2 = \left\|\sum_{j \geq 1} P_{T_j}f\right\|_2 \leq \sum_{j \geq 1}\|P_{T_j}f\|_2 \leq \frac{1}{\sqrt{K}}\sum_{j \geq 0}\|P_{T_j}f\|_1 \leq C\|f\|_2,$$

where the last step uses the hypothesis. Taking $T = T_0$ then gives the result. $\square$

24

What follows is a loose inequality that we will use along with other techniques to prove Theorem 6.

Lemma 6. *Let G be a finite abelian group. Then every $f \in \ell(G)$ satisfies*

$$\|f\|_1 \|\hat{f}\|_1 \geq C\sqrt{N} \|f\|_2^2,$$

where $C = \frac{1}{9}(1 - \frac{1}{\sqrt{2}})$.

Proof. Let $\|f\|_2 = 1$ without loss of generality (the result clearly holds for $f[g] = 0$). First, we quickly rule out the case where $\lfloor 9\|f\|_1^2 \rfloor > N$. Here, we have $\|f\|_1 \geq \sqrt{N}/3$ and $\|\hat{f}\|_1 \geq \|\hat{f}\|_2 = \|f\|_2 = 1$, and so

$$\|f\|_1 \|\hat{f}\|_1 \geq \frac{1}{3}\sqrt{N} \geq C\sqrt{N} \|f\|_2^2,$$

as desired.

For the more interesting case where $\lfloor 9\|f\|_1^2 \rfloor \leq N$, let $T \subseteq G$ denote the indices of the $\lfloor 9\|f\|_1^2 \rfloor$ largest entries of f in absolute value, and $W \subseteq \hat{G}$ the indices of the $\lfloor 9\|f\|_1^2 \rfloor$ largest entries of $\hat{f}$ (this is possible since $\lfloor 9\|f\|_1^2 \rfloor \leq N$). Also, take $\epsilon_T := \|f\|_1 / \sqrt{|T|}$ and $\epsilon_W := \|\hat{f}\|_1 / \sqrt{|W|}$. Then

$$\|f\|_1 = \frac{\|f\|_1}{\sqrt{|T|}} \sqrt{|T|} \|f\|_2 = \epsilon_T \sqrt{|T|} \|f\|_2,$$

and similarly, $\|\hat{f}\|_1 = \epsilon_W \sqrt{|W|} \|\hat{f}\|_2$. As such, Lemma 5 implies that f satisfies the hypothesis of Theorem 2, and so

$$\left(\frac{\|f\|_1}{\epsilon_T} \cdot \frac{\|\hat{f}\|_1}{\epsilon_W} \right)^2 = |T||W| \geq N \big(1 - (\epsilon_T + \epsilon_W) \big)^2.$$

25

We claim that $\epsilon_T, \epsilon_W \in [1/3, 1/\sqrt{8}]$, which implies the result:

$$\|f\|_1^2\|\hat{f}\|_1^2 \geq \frac{1}{81}\left(1 - \frac{1}{\sqrt{2}}\right)^2 N = C^2 N \|f\|_2^4.$$

Rearranging then yields

$$\|f\|_1^2\|\hat{f}\|_1^2 \geq N(\epsilon_T\epsilon_W(1 - (\epsilon_T + \epsilon_W)))^2.$$

To verify the claim, note that $\lfloor 9\|f\|_1^2 \rfloor \leq 9\|f\|_1^2$, and so

$$\epsilon_T = \frac{\|f\|_1}{\sqrt{|T|}} = \frac{\|f\|_1}{\sqrt{\lfloor 9\|f\|_1^2\rfloor}} \geq \frac{\|f\|_1}{\sqrt{9\|f\|_1^2}} = \frac{1}{3}.$$

Also, since $\|f\|_1 \geq \|f\|_2 = 1$, we have $\lfloor 9\|f\|_1^2 \rfloor \geq 9\|f\|_1^2 - 1 \geq 8\|f\|_1^2$, which similarly implies $\epsilon_T \leq 1/\sqrt{8}$. The same logic gives $\epsilon_W \in [1/3, 1/\sqrt{8}]$. $\qquad\square$

Notice that Lemma 6 is a weak version of Theorem 6, but paradoxically, it actually implies Theorem 6. This can be proved using a technique called the *tensor power trick* by Tao [21]. First, we define tensor power.

Definition 4. *For $g_1, \cdots, g_k \in G$ the tensor power of a function $f \in \ell(G)$, denoted $f^{\otimes k}$ is defined as*

$$f^{\otimes k}[g_1, \cdots, g_k] := f[g_1] \cdots f[g_k].$$

where $(g_1, \cdots g_k) \in G^k$.

In order to use $f^{\otimes k}$ and its Fourier transform we need to understand $\widehat{G^k}$. The tensor power trick is only useful if we can show that $\widehat{G^k} = \widehat{G}^k$. Appendix 1.1 gives details that prove this statement. Since $\widehat{G^k} = \widehat{G}^k$, we know $\widehat{f^{\otimes k}} = \hat{f}^{\otimes k}$. With this information, we can prove the following lemmas about $f^{\otimes k}$ and $\hat{f}^{\otimes k}$:

Lemma 7. *Let $p \geq 1$ and $f^{\otimes k}$ be defined as above. Then*

$$\|f^{\otimes k}\|_p = \|f\|_p^k.$$

Proof. We begin with the definition of $\|f^{\otimes k}\|_p^p$.

$$\|f^{\otimes k}\|_p^p = \sum_{(g_1,\cdots g_k)\in G^k} |f^{\otimes k}[g_1,\cdots,g_k]|^p = \sum_{g_1\in G}\cdots\sum_{g_k\in G} |f[g_1]\cdots f[g_k]|^p$$

$$= \sum_{g_1\in G} |f[g1]|^p \cdots \sum_{g_k} |f[g_k]|^p = \left(\|f\|_p^p\right)^k.$$

Taking the pth root of each side yields the result. Therefore, $\|f^{\otimes k}\|_p = \|f\|_p^k.$ $\quad\square$

Lemma 8. *Let $f^{\otimes k}$ be defined as above. Then*

$$\|\widehat{f^{\otimes k}}\|_1 = \|\hat{f}\|_1^k.$$

Proof. By definition of the Fourier transform, and Claims 3 and 5 in Appendix 1.1,

$$\|F_{G^k} f^{\otimes k}\|_1 = \sum_{\chi\in\widehat{G^k}} |(F_{G^k} f^{\otimes k})[\chi]| = \sum_{\chi_1\in\widehat{G}}\cdots\sum_{\chi_k\in\widehat{G}} |(F_G f)[\chi_1]\cdots(F_G f)[\chi_k]|$$

$$= \sum_{\chi_1\in\widehat{G}} |(F_G f)[\chi_1]| \cdots \sum_{\chi_k\in\widehat{G}} |(F_G f)[\chi_k]|.$$

where the last term is the definition of $\|F_G f\|_1^k$. Therefore $\|F_{G^k} f^{\otimes k}\|_1 = \|F_G f\|_1^k.$ $\quad\square$

Now, we are ready to prove Theorem 6 using the tensor power trick.

Proof 3 of Theorem 6. Due to Plancherel's theorem we can rewrite (15) as

$$\|f\|_1 \|\hat{f}\|_1 \geq \sqrt{N} \|f\|_2^2$$

By Lemmas 6, 7 and 8 and we know that

$$\|f^{\otimes k}\|_1 \|\widehat{f^{\otimes k}}\|_1 \geq C\sqrt{N^k}\|f^{\otimes k}\|_2^2$$

implies

$$\|f\|_1^k \|\hat{f}\|_1^k \geq C\left(\sqrt{N}\right)^k \|f\|_2^{2k}.$$

Taking the kth root of both sides yields

$$\|f\|_1 \|\hat{f}\|_1 \geq C^{1/k}\sqrt{N}\|f\|_2^2.$$

Taking the limit as $k \to \infty$ we get $C^{1/k} \to 1$. Thus,

$$\|f\|_1 \|\hat{f}\|_1 \geq \sqrt{N}\|f\|_2^2,$$

completing the proof. □

2.6 Equality in the mixed-norm uncertainty principle

In this section, we will prove a surprising result: that equality in the mixed-norm uncertainty principle is achieved if and only if equality is achieved in the 0-norm uncertainty principle (the "only if" direction is particularly surprising considering Lemma 4). In order to satisfy equality in Theorem 6, the Fourier transform pair f and $\hat{f}$ must have a specific structure:

Theorem 7. *If f satisfies $\mathrm{ns}(f)\,\mathrm{ns}(\hat{f}) = N$, then $f = cM^{\alpha}1_A$ where $c \in \mathbb{C}$, 1_A is the indicator function for some set A, and M^{α} is the modulation operator defined as $(M^{\alpha}f)[n] := f[n]\mathrm{e}^{2\pi\mathrm{i}\alpha n/N}$.*

Proof. We can see from the first two proofs of Theorem 6 that in order to achieve

28

equality in Theorem 6, f and $\hat{f}$ must satisfy equality in Holder's inequality, specifically,

$$\|f\|_1 \|f\|_\infty = \|f\|_2^2, \qquad \|\hat{f}\|_1 \|\hat{f}\|_\infty = \|\hat{f}\|_2^2. \tag{16}$$

To achieve the first equality in (16),

$$\sum_{n \in G} |f[n]|^2 = \|f\|_2^2 = \|f\|_1 \|f\|_\infty = \max_{m \in G} |f[m]| \sum_{n \in G} |f[n]| = \sum_{n \in G} \max_m |f[m]| |f[n]|.$$

Therefore, $\max_{m \in G} |f[m]| = |f[n]|$ for all n such that $f[n] \neq 0$. Similarly, in order for the second equality in (16) to be true, $\max_{m \in \hat{G}} |\hat{f}[m]| = |\hat{f}[n]|$ for all n such that $\hat{f}[n] \neq 0$. The additional constraint for f and $y := \hat{f}$ to satisfy equality in Theorem 6 is equality in the induced norm:

$$\|Ff\|_\infty = \|F\|_{1 \to \infty} \|f\|_1, \qquad \|F^{-1}y\|_\infty = \|F^{-1}\|_{1 \to \infty} \|y\|_1. \tag{17}$$

By definition of the Fourier transform,

$$\|Ff\|_\infty = \max_{m \in \hat{G}} |Ff[m]| = \max_{m \in \hat{G}} \left| \frac{1}{\sqrt{N}} \sum_{n \in G} f[n] e^{-2\pi i m n / N} \right| = \left| \frac{1}{\sqrt{N}} \sum_{n \in G} f[n] e^{-2\pi i \tilde{m} n / N} \right|$$

where $\tilde{m}$ is the index that achieves this maximum. We know that $\|F\|_{1 \to \infty} = 1/\sqrt{N}$. Therefore, we we have

$$\frac{1}{\sqrt{N}} \|f\|_1 = \frac{1}{\sqrt{N}} \sum_{n \in G} |f[n]| = \frac{1}{\sqrt{N}} \sum_{n \in G} |f[n] e^{-2\pi i \tilde{m} n / N}|$$

where the last equality comes from the fact that $|e^{-2\pi i \tilde{m} n / N}| = 1$. Let $v[n] := f[n] e^{-2\pi i \tilde{m} n / N}$. Then we can rewrite the first equality in (17) as

$$\left| \sum_{n \in G} v[n] \right| = \sum_{n \in G} |v[n]|.$$

The statement is true if and only if for all indices n, m such that $v[n] \neq 0$ and $v[m] \neq 0$, $\mathrm{sgn}(v[n]) = \mathrm{sgn}(v[m])$. Thus, $\exists c \in \mathbb{C}$ and $z \geq 0$ entrywise such that $v = cz$. Substituting in the definition of v and solving for f we get

$$f[n] = cz[n]\mathrm{e}^{2\pi i n \tilde{m}/N}.$$

Therefore, to satisfy equality in Theorem 6, f and $\hat{f}$ have the form $f = cM^{\alpha}1_A$. $\quad\square$

Theorem 7 above can be used to prove the main result of this section:

Theorem 8. $\mathrm{ns}(f)\,\mathrm{ns}(\hat{f}) = N$ *if and only if* $\|f\|_0\|\hat{f}\|_0 = N$.

Proof. By Lemma 4 and Theorem 1, we know that

$$\|f\|_0\|\hat{f}\|_0 \geq \mathrm{ns}(f)\,\mathrm{ns}(\hat{f}) \geq N.$$

Therefore, functions which satisfy equality for Theorem 6 form a subset of the functions that achieve equality for Theorem 1.

Assume that f satisfies equality in Theorem 6. Then using Theorem 7, the numerical sparsity of f can be calculated as follows:

$$\mathrm{ns}(f) = \frac{\|f\|_1^2}{\|f\|_2^2} = \frac{\left(\sum_{n=1}^{N}|f[n]|\right)^2}{\|f\|_\infty\|f\|_1} \frac{\left(\sum_{n=1}^{N}|CM^{\alpha}1_A[n]|\right)^2}{C \cdot \sum_{n=1}^{N}|CM^{\alpha}1_A[n]|} = \frac{\|f\|_0^2 \cdot C^2}{\|f\|_0 \cdot C^2} = \|f\|_0.$$

Thus, the set of functions that achieve equality in Theorem 6 is the same set that achieves equality in Theorem 1. $\quad\square$

Consider the case when K divides N and f is a Dirac comb supported on multiples of K. It is well known that $\hat{f}$ is a Dirac comb supported on multiples of N/K, and so $\|f\|_0\|\hat{f}\|_0 = N$. Thus, by Theorem 8, f achieves equality in Theorem 6 in this special case. As such, it is possible to nontrivially achieve equality in Theorem 6 if and only if N is composite (the "only if" direction follows from the fact that $\|f\|_0$ and $\|\hat{f}\|_0$ are both integers). In the next section, we give an example of a function that achieves near equality in Theorem 6 regardless of N.

2.7 Near equality in the mixed-norm uncertainty principle

In this section, we will show that a discretized and periodized Gaussian function achieves near equality in the mixed-norm uncertainty principle in the following sense:

Theorem 9. *There exists a $C > 0$ independent of N and $f \in \ell(\mathbb{Z}_N)$ such that*

$$\operatorname{ns}(f) \operatorname{ns}(\hat{f}) \leq CN.$$

Once again, let $f = h^{(K)}$ as defined in (11) with $K = \sqrt{N}$. Since $\hat{f} = f$, it suffices to show that $\operatorname{ns}(f) \leq \sqrt{CN}$. A proof of this statement requires the use of the following lemma:

Lemma 9. *Given $h^{(K)}$ where $K = \sqrt{N}$, $\|h^{(K)}\|_1 \leq \frac{3}{2}N^{3/4}$.*

Proof. By definition,

$$\|h^{(K)}\|_1 = \sum_{n=1}^{N} |h^{(K)}[n]| = \sum_{n=1}^{N} N^{\frac{1}{4}} \left| \sum_{n' \in \mathbb{Z}} \exp\left[-\pi \left(\frac{n}{\sqrt{N}} - n'\sqrt{N}\right)^2\right] \right|.$$

Because $\exp(y) \geq 0$ for all $y \in \mathbb{R}$ the above we then have

$$\|h^{(K)}\|_1 = \sum_{n=1}^{N} N^{\frac{1}{4}} \sum_{n' \in \mathbb{Z}} \exp\left[-\pi \left(\frac{n}{\sqrt{N}} - n'\sqrt{N}\right)^2\right]. \tag{18}$$

The following is an upper bound on (18) due to integral comparison.

$$N^{\frac{1}{4}} \sum_{n=1}^{N} \int_{-\infty}^{\infty} \exp\left[-\pi \left(\frac{n}{\sqrt{N}} - x\sqrt{N}\right)^2\right] dx + N^{\frac{1}{4}} \sum_{n=1}^{N} \exp\left[-\pi \left(\frac{n}{\sqrt{N}}\right)^2\right]. \tag{19}$$

31

First we will find an upper bound on the left hand side of (19). Substituting $u = \sqrt{\pi}(n/\sqrt{N} + x\sqrt{N})$ into the first term of (19) gives

$$N^{\frac{1}{4}} \sum_{n=1}^{N} \frac{1}{\sqrt{\pi N}} \int_{-\infty}^{\infty} \exp[-u^2]du + N^{5/4} = N^{\frac{1}{4}} \sum_{n=1}^{N} \frac{\sqrt{\pi}}{\sqrt{\pi N}} = N^{\frac{3}{4}}.$$

The upper bound on the second term of (19) is more difficult to compute. By symmetry,

$$N^{\frac{1}{4}} \sum_{n=1}^{N} \exp\left[-\pi\left(\frac{n}{\sqrt{N}}\right)\right] = N^{\frac{1}{4}} \sum_{n=-N}^{-1} \exp\left[-\pi\left(\frac{n}{\sqrt{N}}\right)\right] \leq \int_{-N}^{-1} \exp\left[-\pi\frac{x^2}{N}\right] dx.$$

We know that

$$\int_{-N}^{-1} \exp\left[-\pi\frac{x^2}{N}\right] dx \leq \int_{-\infty}^{0} \exp\left[-\pi\frac{x^2}{N}\right] dx = \sqrt{\frac{N}{\pi}} \int_{-\infty}^{0} \exp[-u^2]du = \frac{\sqrt{N}}{2},$$

where $u = \sqrt{\pi/N}x$. Therefore, the upper bound on the second term of (19) is $N^{3/4}/2$. Combining this upper bound with the one for the first term yields the result. $\qquad\square$

Proof of Theorem 9. It is straightforward to construct our upper bound on $\mathrm{ns}(h^{(K)})$ using a combination of Lemma 9 and Claim 2 (see Appendix 1.1):

$$\mathrm{ns}(h^{(K)}) \leq \frac{9}{4} \frac{N^{3/2}}{\frac{N}{\sqrt{2}} - \sqrt{N}} = \frac{9}{4} \frac{N}{\sqrt{\frac{N}{2}} - 1} = \frac{9}{4}\sqrt{2}\frac{N}{\sqrt{N} - \sqrt{2}} = \frac{9\sqrt{2}}{4}(1 + o_N(1))\sqrt{N}.$$

(20)

Thus, for some $C > 0$ we have $\mathrm{ns}(f)\,\mathrm{ns}(\hat{f}) \leq CN$. $\qquad\square$

III. Applications of the uncertainty principle

In Chapter II, we introduced a mixed-norm uncertainty principle and demonstrated that near equality is achieved with a discretized and periodized Gaussian. In this chapter, we will discuss two applications of these results. First, we will show a fundamental limitation of the demixing problem, namely, that demixing N-dimensional K-sparse signals is only stably possible in the worst case if $K = O(\sqrt{N})$. Second, we will show how to detect K-sparse signals with only $O(K)$ measurements.

3.1 Limitations of the demixing problem

The main idea of demixing is that if a signal x which is sparse in the Fourier domain is corrupted with noise ϵ that is sparse in the time domain, then we can use compressed sensing methods to reconstruct the original signal x given the corrupted signal $z = x + \epsilon$. This reconstruction is done by solving

$$v^\star = \operatorname{argmin} \|v\|_1 \qquad \text{subject to } [I\ F]v = Fz,$$

where the solution $v^\star$ is a vertical concatenation of Fx and ϵ.

Coherence-based guarantees in [22] show that it suffices for $v^\star$ to be K-sparse with $K = O(\sqrt{N})$ while RIP-based guarantees in [6] allow for $K = O(N)$ if $[I\ F]$ is replaced with a random matrix. We refer to this disparity as the *square-root bottleneck*. In particular, does $[I\ F]$ perform similarly to a random matrix or is the coherence-based sufficient condition on K also necessary? Despite being studied in both [10] and [18], this fundamental problem has gone unsolved for general N until now. In this section, we use numerical sparsity to show that $\Phi = [I\ F]$ cannot break the square-root bottleneck. We begin with a definition:

Definition 5. *Let $\Phi \in \mathbb{C}^{M \times N}$. We say that Φ satisfies the C-width property if*

$$\|x\|_2 \leq \frac{C}{\sqrt{K}} \|x\|_1 \qquad \forall x \in \mathrm{Ker}(\Phi).$$

The following theorem is provided in [23] and [24]:

Theorem 10. *Define*

$$\Delta(y) := \mathrm{argmin} \, \|x\|_1 \quad \text{such that } \Phi x = y.$$

Then $\exists C > 0$ such that

$$\|\Delta(\Phi x) - x\|_2 \leq \frac{C}{\sqrt{K}} \|x - x_K\|_1 \quad \forall x \in \mathbb{R}^N$$

if and only if $\exists c > 0$ such that

$$\|x\|_2 \leq \frac{c}{\sqrt{K}} \|x\|_1 \qquad \forall x \in \mathrm{Ker}(\Phi).$$

Furthermore, $C \asymp c$ in both directions of the equivalence.

Based on Definition 3, the C-width property is equivalent to having

$$\mathrm{ns}(x) \geq K/C^2 \qquad \forall x \in \mathrm{Ker}(\Phi).$$

Let $\Phi = [I \; F]$ and $h^{(K)}$ be defined as in (11). Take $x \in \mathbb{C}^{2N}$ such that

$$x = \begin{bmatrix} h^{(K)} \\ -h^{(K)} \end{bmatrix} \tag{21}$$

for $K = \sqrt{N}$. We know that $x \in \mathrm{Ker}\,\Phi$. We will show that $\mathrm{ns}(x) \leq C\sqrt{M}$. This is significant because by Theorem 10, it provides a converse to the square-root bottleneck for this well-studied matrix. The proof of this statement requires the use of the following lemma:

Lemma 10. *For x as defined in (21), $\mathrm{ns}(x) = 2\,\mathrm{ns}(h^{(K)})$.*

Proof. We will first show that $\|x\|_1^2 = 4\|h^{(K)}\|_1^2$. The definition of x and $h^{(K)}$ gives

$$
\|x\|_1 = \sum_{i=1}^{2N} |x_i|
$$

$$
= \sum_{i=1}^{N} |x_i| + \sum_{i=N+1}^{2N} |x_i|
$$

$$
= \sum_{i=1}^{N} |h^{(K)}[i]| + \sum_{i=1}^{N} |-h^{(K)}[i]| = 2\sum_{i=1}^{N} |h^{(K)}[i]| = 2\|h^{(K)}\|_1.
$$

Squaring the above yields $\|x\|_1^2 = 4\|h^{(K)}\|_1^2$. Next, we will show that $\|x\|_2^2 = 2\|h^{(K)}\|_2^2$. Similar to the above by definition,

$$
\|x\|_2^2 = \sum_{n=1}^{2N} |x_n|^2
$$

$$
= \sum_{n=1}^{N} |x_n|^2 + \sum_{n=N+1}^{2N} |x_n|^2
$$

$$
= \sum_{n=1}^{N} |h^{(K)}[n]|^2 + \sum_{n=1}^{N} |-h^{(K)}[n]|^2 = 2\sum_{n=1}^{N} |h^{(K)}[n]|^2 = 2\|h^{(K)}\|_2^2.
$$

Combining the above results gives us

$$
\mathrm{ns}(x) = \frac{\|x\|_1^2}{\|x\|_2^2} = \frac{4\|h^{(K)}\|1^2}{2\|h^{(K)}\|2^2} = 2\,\mathrm{ns}(h^{(K)}),
$$

completing the proof $\qquad\square$

Therefore, based on Lemma 10 and (20) we have

$$\mathrm{ns}(x) \le \frac{9\sqrt{2}}{2} \frac{N}{\sqrt{N} - \sqrt{2}} = \frac{9}{\sqrt{2}} \left(1 + o_N(1)\right) \sqrt{N},$$

as desired.

3.2   Fast detection of sparse signals

In this section, we will show that with high probability, an arbitrary K-sparse signal can be detected using $O(K)$ measurements. We want to be able to distinguish between a sparse signal and the zero signal. Therefore, we are testing the following hypotheses:

$$H_0 : x = 0$$
$$H_1 : \|x\|_2^2 = \frac{N}{K}, \|x\|_0 \le K.$$

Here, we assume we know the 2-norm of the sparse vector we intend to detect, and we set it to be $\sqrt{N/K}$ without loss of generality (this choice of scaling will help us interpret our results later). We will assume the data is accessed according to the following query-response model:

Definition 6 (Query-response model). *If the ith query is $n_i \in \mathbb{Z}_N$, then the ith response is $(Fx)[n_i] + \epsilon_i$, where the ϵ_i's are iid complex random variables with some distribution such that*

$$\mathbb{E}|\epsilon_i| = \alpha, \qquad \mathbb{E}|\epsilon_i|^2 = \beta^2.$$

The coefficient of variation v of $|\epsilon_i|$ is defined as

$$v^2 = \frac{\mathrm{Var}\,|\epsilon_i|}{(\mathbb{E}|\epsilon_i|)^2} = \frac{\beta^2 - \alpha^2}{\alpha^2}.$$

Note that for any scalar $c \neq 0$, the mean and variance of $|c\epsilon_i|$ are $|c|\alpha$ and $|c|^2 \operatorname{Var} |\epsilon_i|$, respectively. Hence, v is scale invariant and is simply a quantification of the "shape" of the distribution of $|\epsilon_i|$. We will evaluate the responses to our queries with an ℓ_1-detector, defined below.

Definition 7 (ℓ_1-detector). *Fix a threshold τ. Given responses $\{y_i\}_{i=1}^M$ from the query-response model, if*

$$\sum_{i=1}^M |y_i| > \tau,$$

then we reject H_0.

The following is the main result of this section:

Theorem 11. *Suppose $\alpha \leq 1/(8K)$. Randomly draw M independent indices uniformly from $\mathbb{Z}_N$, input them into the query-response model and apply the ℓ_1-detector with threshold $\tau = 2M\alpha$ to the responses. Then*

$$\Pr\left(\text{reject } H_0 \middle| H_0 \right) \leq p$$

and

$$\Pr\left(\text{fail to reject } H_0 \middle| H_1 \right) \leq p + q$$

provided $M \geq 4K/q + v^2/p$.

In words, the probability that the ℓ_1-detector delivers a false positive is p and the probability that it delivers a false negative is $p + q$. These error probabilities can be estimated better given more information about the distribution of the random noise, and the threshold τ can be modified to decrease one error probability at the price of increasing the other. Notice that we only use $O(K)$ samples in the Fourier domain to detect a K-sparse signal. Since the sampled indices are random, it will take $O(\log N)$ bits to communicate each query, leading to a total computational burden of $O(K \log N)$ operations. We suspect K-sparse signals cannot be detected with substantially fewer samples (in the Fourier domain or any domain).

We also note that the acceptable noise magnitude $\alpha = O(1/K)$ is optimal in some sense. To see this, consider the case where K divides N and x is a Dirac comb of K deltas. Then Fx is a Dirac comb of N/K deltas. (Thanks to our choice of scaling, each delta in the Fourier domain has unit magnitude.) Since a proportion of $1/K$ entries is nonzero in the Fourier domain, we can expect to require $O(K)$ random samples in order to observe a nonzero entry, and the ℓ_1-detector will not distinguish the entry from accumulated noise unless $\alpha = O(1/K)$. We will use the following lemmas to prove Theorem 11.

Lemma 11. *Suppose $\|x\|_0 \leq K$ and $\|x\|_2^2 = N/K$. Draw $n \sim \text{Unif}(\mathbb{Z}_N)$ and define $Y := |(Fx)[n]|$. Then*

$$\mathbb{E}Y \geq \frac{1}{K}, \qquad \mathbb{E}Y^2 = \frac{1}{K}.$$

Proof. By Lemma 4 we know $\text{ns}(x) \leq \|x\|_0 \leq K$, and Theorem 6 gives

$$N \leq \text{ns}(x)\,\text{ns}(Fx) \leq K\,\text{ns}(Fx).$$

Rearranging and the definition of numerical sparsity then gives

$$\frac{N}{K} \leq \text{ns}(Fx) = \frac{\|Fx\|_1^2}{\|Fx\|_2^2} = \frac{\|Fx\|_1^2}{\|x\|_2^2} = \frac{\|Fx\|_1^2}{N/K}$$

where the second to last equality is due to Plancherel's Theorem. Thus $\|Fx\|_1 \geq N/K$. Then

$$\mathbb{E}Y = \frac{1}{N}\sum_{n \in \mathbb{Z}_N} Y_n = \frac{1}{N}\sum_{n \in \mathbb{Z}_N} |(Fx)[n]| = \frac{1}{N}\|Fx\|_1 \geq \frac{1}{K}$$

and

$$\mathbb{E}Y^2 = \frac{1}{N}\sum_{n\in\mathbb{Z}_N} Y_n^2 = \frac{1}{N}\sum_{n\in\mathbb{Z}_n} |(Fx)[n]|^2 = \frac{1}{N}\|Fx\|_2^2 = \frac{1}{K}. \qquad \square$$

Lemma 12. *Take $\epsilon_1,\ldots,\epsilon_M$ to be iid complex random variables with $\mathbb{E}|\epsilon_i| = \alpha$ and $\mathbb{E}|\epsilon_i|^2 = \beta^2$. Then*

$$\Pr\left(\sum_{i=1}^{M}|\epsilon_i| \leq M\alpha\right) \geq 1-p$$

provided $M \geq v^2/p$, where v is the coefficient of variation of $|\epsilon_i|$.

Proof. Chebyshev's inequality gives

$$\Pr\left(|X - \mathbb{E}X| > t\right) \leq \frac{\operatorname{Var} X}{t^2}.$$

Take $X = \sum_{i=1}^{M}|\epsilon_i|$. Then

$$\mathbb{E}X = \sum_{i=1}^{M}\mathbb{E}|\epsilon_i| = M\mathbb{E}|\epsilon_i| = M\alpha$$

and

$$\operatorname{Var} X = M\operatorname{Var}|\epsilon_i| = M(\mathbb{E}|\epsilon_i|^2 - (\mathbb{E}|\epsilon_i|)^2) = M(\beta^2 - \alpha^2).$$

Thus,

$$\Pr\left(\sum_{i=1}^{M}|\epsilon_i| > M\alpha + t\right) = \Pr\left(\sum_{i=1}^{M}|\epsilon_i| - M\alpha > t\right)$$
$$\leq \Pr(|X - \mathbb{E}X| > t) \leq \frac{\operatorname{Var} X}{t^2} = \frac{M(\beta^2 - \alpha^2)}{t^2}.$$

Take $t = M\alpha$. Then

$$\Pr\left(\sum_{i=1}^{M}|\epsilon_i| > 2M\alpha\right) \leq M\frac{(\beta^2 - \alpha^2)}{(M\alpha)^2} = \frac{\beta^2 - \alpha^2}{M\alpha^2} \leq \frac{\beta^2 - \alpha^2}{\alpha^2}\cdot\frac{p}{v^2} = p.$$

which gives the result. $\qquad\square$

Lemma 13. *If* $M \geq 4K/q + v^2/p$, *and* $a \leq 1/(8K)$, *then*

$$\Pr\left(\sum_{i=1}^{M} |y_i| < \tau \,\middle|\, \|x\|_0 \leq K, \|x\|_2^2 = \frac{N}{K}\right) = p + q.$$

Proof. We know that

$$\Pr\left(\sum_{i=1}^{M} |y_i| \leq 2Ma\right) \leq \Pr\left(\sum_{i=1}^{M} Y_i - \sum_{i=1}^{M} |\epsilon_i| \leq 2Ma\right). \tag{22}$$

For simplicity let $b = \sum_{i=1}^{M} Y_i - \sum_{i=1}^{M} |\epsilon_i|$. By Lemma 12, we then have

$$\Pr(b \leq 2M\alpha) = \Pr\left(b \leq 2M\alpha \,\middle|\, \sum_{i=1}^{M} |\epsilon_i| > 2M\alpha\right) \Pr\left(\sum_{i=1}^{M} |\epsilon_i| > 2M\alpha\right)$$

$$+ \Pr\left(b \leq 2M\alpha \,\middle|\, \sum_{i=1}^{M} |\epsilon_i| \leq 2M\alpha\right) \Pr\left(\sum_{i=1}^{M} |\epsilon_i| \leq 2M\alpha\right)$$

$$\leq p + \Pr\left(\sum_{i=1}^{M} Y_i \leq 4M\alpha\right). \tag{23}$$

Chebyshev's inequality gives

$$\Pr(|X - \mathbb{E}X| > t) \leq \frac{\operatorname{Var} X}{t^2}.$$

Taking $X = \sum_{i=1}^{M} Y_i$, Lemma 11 gives $\mathbb{E}X = M\mathbb{E}Y_i \geq \frac{M}{K}$ and $\operatorname{Var} X = M \operatorname{Var} Y_i$. Thus,

$$\Pr\left(\sum_{i=1}^{M} Y_i < \frac{M}{K} - t\right) \leq \Pr(X \leq \mathbb{E}X - t) \leq \Pr(|X - \mathbb{E}X| > t) \leq \frac{\operatorname{Var}(X)}{t^2} \leq \frac{M}{Kt^2}.$$

Taking $t = M/(2K)$ gives

$$\Pr\left(\sum_{i=1}^{M} Y_i \leq 4M\alpha\right) \leq \Pr\left(\sum_{i=1}^{M} Y_i \leq \frac{M}{2K}\right)$$

$$= \Pr\left(\sum_{i=1}^{M} Y_i \leq \frac{M}{K} - t\right)$$

$$\leq \frac{M}{Kt^2} = \frac{M}{K}\left(\frac{2K}{M}\right)^2 = \frac{4K}{M} < q. \tag{24}$$

Combining (22), (23), and (24) gives the result. $\qquad\qquad\qquad\square$

Thus, Lemma 13 and 12 together prove Theorem 11.

IV. Fast hyperspectral imaging using compressed sensing

Astronomers have been attempting to analyze the spectra of stars for over a hundred years [25]. With modern technology, scientists can look at hyperspectral images of stars and use the data to determine their chemical makeup. This concept is known as *spectroscopy*. The main phenomena of interest are absorption lines in the spectral bands. Absorption lines are a significant drop in the intensity of the light at a given frequency, or spectral band, compared to the distribution of the surrounding spectral bands (see Figure 6). The locations of these absorption lines can be matched with the light emitted from super heating particular elements of the periodic table, which then allows scientists to determine the chemical composition of a star.

Conventional hyperspectral cameras are slow. Different methods of hyperspectral imaging either require time to process the entire space desired or greatly limit the number of spectral bands [27]. Before describing these methods, it is helpful to introduce some notation. Let $X = f(x, y, \lambda)$ be a function representing the hyperspectral image we wish to measure. This data cube is made up of two spatial dimensions with coordinates represented by x and y and one spectral dimension whose coordinate is represented by λ. In particular, for any fixed spatial coordinates (x, y), varying λ gives the spectrum of light received at (x, y).

There are two conventional methods for hyperspectral imaging: spectral scanning and spatial scanning. Spectral scanning involves the use of filters, such as bandpass filters, to collect a two-dimensional slice of X that contains only one fixed spectral band λ and all spatial information. Filters are commonly installed on wheels that rotate in front of the lens, allowing only one spectral band to be captured at a time [13, 27]. Thus, time is needed to collect all spectral bands in X.

There are two types of spatial scanning methods. Point-scanning (or whisk-broom) methods rely on moving mechanical parts in the camera. This method uses a mirror

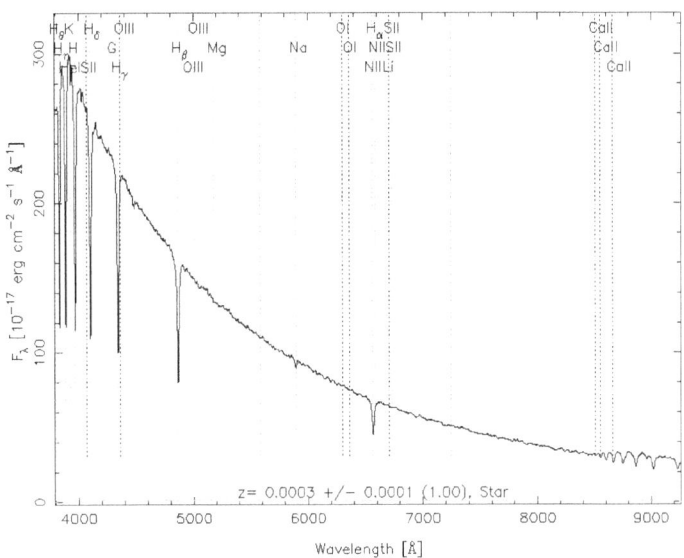

Figure 6. Spectrum of a star (object identification number 587722953303982115) from the Sloan Digital Sky Survey [26]. Here, we see the spectrum of this star over a wavelength range of 400-900 nm. As such, this data includes both visible and infrared light. Though the data is noisy, there are at least seven distinct absorption lines throughout the spectra.

to scan across all x coordinates of X for each y coordinate. The mirror then reflects the light from a fixed point in space through a pinhole and then a prism that disperses the light before it is recorded by charge-coupled device (CCD). Therefore, spectral information of only one point in space is captured by the camera at a time. The need to scan in both spatial directions is time consuming and requires complex hardware that is prone to degrade over time. On the other hand, whisk-broom scanning has the advantage of only requiring one detector to calibrate [13, 27].

The line-scanning (or push-broom) method involves passing light through a vertical slit so that all spectral bands of the desired image are collected for some fixed spatial coordinate x. Therefore, a two-dimensional slice of the data cube is collected (with one spatial and one spectral dimension). The light is passed through a prism that disperses the different frequencies before it is collected by a CCD [13, 27]. The major limitations of the spatial scanning methods are (1) the amount of time it takes

43

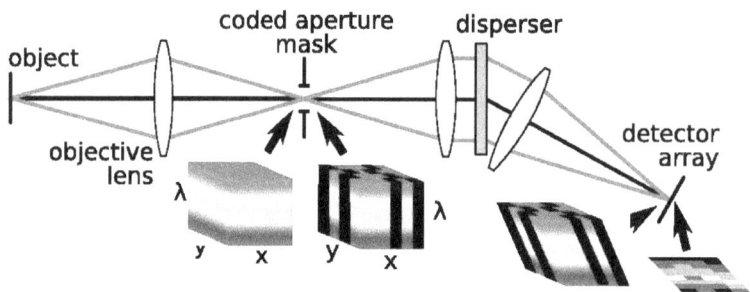

Figure 7. Schematic for the coded aperture snapshot spectral imager [27]. The coded aperture mask in question is a micro-mirror array and a prism is used as the disperser. In this chapter, we consider a similar sensing platform, which we model in Section 4.1

for the long exposure required to image each portion of the data cube and (2) the necessity to move the camera in order to capture each portion.

Luckily, the distribution of stars throughout space is highly sparse. Therefore, instead of using time-consuming scanning methods to collect hyperspectral data, we can apply compressed sensing methods. The mechanism we will use is similar to other spectroscopy mechanisms in that it involves the use of a prism or some other dispersion device before the data is collected on a CCD.

Unlike other spectroscopy platforms, this platform makes use of a micro-mirror array (MMA) [15], that is, a rectangular grid of mirrors, each corresponding to a spatial location (x, y). Each mirror is oriented either to reflect all available light at (x, y) through the prism onto the CCD or to reflect that light away from the prism and CCD so that it is not measured (see Figures 4 and 7). Therefore, instead of measuring the data cube one slice at a time, we measure various combinations of data cube entries. We intend to exploit the data cube's sparsity in order to reconstruct it from these measurements.

Other methods of compressive hyperspectral imaging have been developed recently. One method is called *chromotomography*, which is a generalized scanning method that does not require spatial filtering such as the slit used in line scanning methods. Rather, chromotomography involves a rotating prism [11, 27]. A disadvantage of this method is that the dispersion elements required are difficult to produce [27]. Another method is called *coded aperture imaging*, which is a snapshot method that uses coded apertures (e.g., MMA) in place of other spatial filtering devices used for scanning methods [28, 29]. The model we propose is similar to the coded aperture systems developed in [28, 29].

4.1 Modeling the sensor

Let $X = f(x, y, \lambda)$ be a data cube like that in Figure 3 with spatial dimensions of sizes J and L and with N spectral bands, that is, $X \in \mathbb{R}^{J \times L \times N}$. The entire measurement platform (i.e., the MMA, prism, and CCD) can be represented as a linear operator A_v, where v represents the orientation of the MMA's mirrors. We will construct A_v in terms of two operators Φ_v and Ψ such that $A_v = \Psi \Phi_v$, where Φ_v is a model of the MMA and Ψ is a model of the prism and CCD combined. As mentioned in the previous section, the MMA either reflects all spectral bands for a given point (x, y) towards the prism and CCD or reflects all spectral bands away from the CCD. As such, the MMA transforms the data cube according to a linear operator $\Phi_v : \mathbb{R}^{J \times L \times N} \to \mathbb{R}^{J \times L \times N}$ defined by

$$\Phi_v \delta_{(x,y,\lambda)} := v_{(x,y)} \delta_{(x,y,\lambda)} \tag{25}$$

where $v_{(x,y)} = 1$ if the mirror at (x, y) allows light through and is otherwise zero. Next, the prism disperses the different frequencies of light by different amounts before the CCD records light intensities. We model this by normalizing x, y and λ so that $x \in \{1, \ldots, J\}$, $y \in \{1, \ldots, L\}$ and $\lambda \in \{1, \ldots, N\}$, and then modeling the dispersion process with a linear operator $\Psi : \mathbb{R}^{J \times L \times N} \to \mathbb{R}^{J \times (L+N-1)}$ defined by

$$\Psi \delta_{(x,y,\lambda)} := \delta_{(x,y+\lambda-1)}. \tag{26}$$

That is, light at (x,y) of wavelength λ contributes to the $(x, y + \lambda + 1)$th entry of the CCD. Indeed, in our model, the prism is oriented so that dispersion acts in the y direction. At this point, we note that $(\Psi \Phi_v X)[x, \cdot]$ is completely determined by $X[x, \cdot, \cdot]$, meaning we may simplify our analysis by considering these slices of X in parallel. As such, for the remainder of this thesis, we will assume $J = 1$ without loss of generality.

Let's briefly describe the matrix representation of A_v. We choose the identity basis for both $\mathbb{R}^{1 \times L \times N}$ and $\mathbb{R}^{1 \times (L+N-1)}$. Specifically, we order the basis elements of $\mathbb{R}^{1 \times L \times N}$ as $\delta_{(1,1,1)}, \delta_{(1,1,2)}, \ldots, \delta_{(1,1,N)}, \delta_{(1,2,1)}, \ldots, \delta_{(1,L,N)}$. Then z is a vector of coefficients of X in this basis. By (25), Φ_v is a multiplication operator that only depends on y. As such, its matrix representation in this basis is a diagonal matrix composed of L diagonal blocks, where the yth block is $v_{(1,y)} I_{N \times N}$. By (26), Ψ is a translation and sum operator that depends on y and λ. The matrix representation uses the translation operator T, where

$$T^{y-1} \delta_\lambda := \delta_{y+\lambda-1}$$

for identity basis elements $\delta_\lambda \in \ell(\mathbb{Z}_{L+N-1})$. We apply this cyclic translation with the help of a zero-padding matrix:

$$B := \begin{bmatrix} I_{N \times N} \\ 0_{L-1 \times N} \end{bmatrix}.$$

We then write

$$\Psi := [B \ \ T^1 B \ \ \cdots \ \ T^{L-1} B]. \tag{27}$$

Overall, we have the measurement matrix $A_v = \Psi\Phi_v$ when the MMA mirrors are oriented according to v. Given multiple exposures with varying MMA orientations $\{v_i\}_{i=1}^{Q}$, our observations can be organized as $y = Az$, where

$$A = \begin{bmatrix} A_{v_1} \\ \vdots \\ A_{v_Q} \end{bmatrix}. \tag{28}$$

In total, A has $M := Q(N + L - 1)$ rows and $P := LN$ columns.

4.2 Modeling the data cube

Recall from the previous section that the data cube is $1 \times L \times N$ without loss of generality, and we reshape this cube to form the vector $z \in \mathbb{R}^{LN} = \mathbb{R}^P$. In particular, the first length-N block of z corresponds to the spectrum with spatial coordinates $(1,1)$ and the second length-N block corresponds to $(1,2)$, etc. Since the scene will only have a few stars, we have that z is sparse, with the yth block giving the spectrum of the star at $(1,y)$ unless no such star exists, in which case the yth block is identically zero. Therefore, the nonzero entries of z are clustered into blocks, and we say that z is *block sparse*, as defined below.

Definition 8 (Block sparsity). *Let z be a vector of length P such that z is a concatenation of blocks of length N. We say that z is K-block sparse if at most K blocks in z are nonzero. We denote the nth block of z by z_n.*

When a star is present in the data, we model the spectral radiance of the star as black-body radiation. As such, we use Planck's Law as part of our model of the nonzero blocks of z [30].

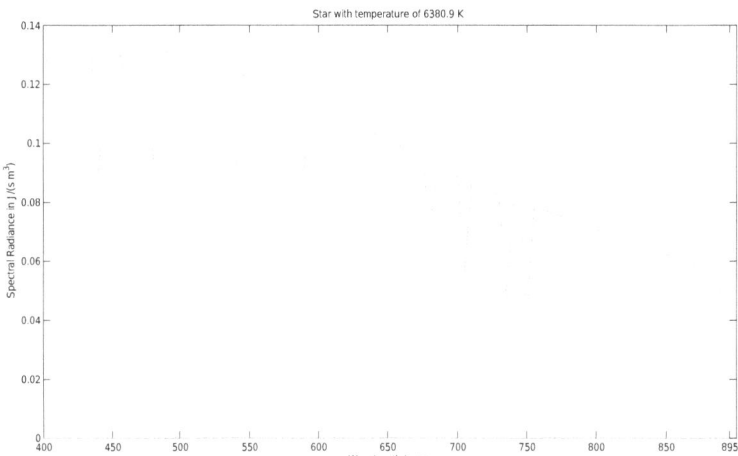

Figure 8. Example of star simulated in MATLAB with temperature of 6380.9 K and ten absorption lines using `stardata.m` (See Appendix B). This curve closely resembles the real-world data in Figure 6.

Definition 9 (Planck's Law). *The spectral radiance of a black body is a function of wavelength and temperature:*

$$L(\lambda, T) = \frac{2hc^2}{\lambda^5} \left[\exp\left(\frac{hc}{\lambda k T}\right) - 1 \right]^{-1},$$

where c, h and k are the speed of light, the Planck constant, and the Boltzmann constant, respectively. Additionally, λ is wavelength in meters and T is temperature in Kelvin.

As mentioned in the beginning of this chapter, absorption lines in the spectral radiance of stars indicate the elements that are present in said star. We can identify what elements are present because these absorption lines are the negative of the emission lines generated when those elements are heated to extreme temperatures [31]. In the absence of interference, these lines exhibit Lorentzian shapes [32, 33]. For each star, we model s absorption lines in the spectral data as negative Lorentzian functions.

In particular, an absorption line at wavelength λ_i is given by

$$H_i(\lambda) := -\frac{\alpha}{1 + (\lambda - \lambda_i)^2/\beta}, \tag{29}$$

where α and β are shape parameters. Data collected for a star with absorption lines at $\lambda_1, \ldots, \lambda_s$ is modeled as $L(\lambda, T) + \sum_{k=1}^{s} H_k(\lambda)$. Figure 8 illustrates an example of this simulated data.

4.3 Block orthogonal matching pursuit

In this section, we define the block orthogonal matching pursuit (BOMP) algorithm and conditions for guaranteed reconstruction of the desired data. Take $A \in \mathbb{R}^{M \times P}$ and let $z \in \mathbb{R}^P$ be a K-block sparse vector with blocks of size $N = P/L$. BOMP is an iterative greedy algorithm that recovers z from $w = Az$. Let A_i be the ith $M \times N$ block of A and z_j be the jth block of z. Initialize $\mathcal{I}^0$ to be the empty set, the residual $r^0 = w$, and z^0 as the zero vector. On the lth iteration of BOMP, an additional element of the support $\mathcal{I}$ is found:

$$\mathcal{I}^l = \mathcal{I}^{l-1} \cup \left\{ \arg\max_{i \in \mathbb{Z}_N} \|A_i^* r^{l-1}\|_2 \right\}. \tag{30}$$

Then given $\mathcal{I}^l$, we solve for z^l using least-squares minimization:

$$z^l = \arg\min_{\tilde{z}} \left\| w - \sum_{i \in \mathcal{I}^l} A_i \tilde{z}_i \right\|_2. \tag{31}$$

Lastly, we update the residual:

$$r^l = w - \sum_{i \in \mathcal{I}^l} A_i z_i^l. \tag{32}$$

In the case of traditional sparsity, if a matrix A has small coherence, defined below, it has been shown that A can be used to successfully recover any sparse signal.

Definition 10 (Coherence). *For a matrix A, define the coherence of A, $\mu(A)$ as*

$$\mu(A) := \max_{i,j \neq i} |a_i^* a_j|$$

Specifically, it is known that orthogonal matching pursuit will recover any K-sparse z from $w = Az$ provided $\mu > (2K - 1)^{-1}$ [9]. Eldar et al. [16] developed two analogous coherence definitions that lead to a similar guarantee for block orthogonal matching pursuit. They define block-coherence as follows:

Definition 11 (Block-coherence). *For $A = [A_1 \cdots A_L]$ with $M \times N$ blocks A_k, we define the block-coherence of A to be*

$$\mu_B(A) := \max_{i,j \neq i} \frac{1}{N} \|A_i^* A_j\|_{2 \to 2}$$

Notice that Definition 11 does not account for pairs columns within a common block of A. These columns are considered by sub-coherence:

Definition 12 (Sub-coherence). *For $A = [A_1 \cdots A_L]$ with $M \times N$ blocks A_k, we define the sub-coherence of A to be*

$$\nu(A) = \max_k \max_{i,j \neq i} |(a_i^{(k)})^* a_j^{(k)}|,$$

where $a_i^{(k)}$ and $a_j^{(k)}$ are columns of A_k.

Definitions 11 and 12 lead to a block version of the coherence condition [16]:

Theorem 12. *Let $z \in \mathbb{C}^P$ be a K-block sparse vector with block length $N = P/L$ and $w = Az$ for $A \in \mathbb{C}^{M \times P}$ with columns of unit 2-norm. For BOMP to recover z, it*

50

suffices that

$$KN < \frac{1}{2}\left(\frac{1}{\mu_B(A)} + N - (N-1)\frac{\nu(A)}{\mu_B(A)}\right).$$

Furthermore, BOMP converges to z in at most K steps, a vast improvement on the KN steps that would be required with OMP.

4.4   Performance guarantee for fast hyperspectral imaging

In this section, we will show that the sensor model in Section 4.1 satisfies the conditions in Theorem 12 for recovery using block orthogonal matching pursuit (BOMP). We begin with a definition:

Definition 13. *A matrix $V \in \{0,1\}^{Q \times L}$ is said to be an (R, γ)-incoherent design if*

(i) every column of V has exactly R ones.

(ii) every pair of columns has at most γ ones in common.

The following is the main result of this section:

Theorem 13. *Take A as defined in (28), where each v_q is given by a different row of an (R, γ)-incoherent design. The BOMP algorithm will recover any K-block sparse z from $w = Az$ provided*

$$K < \frac{1}{2}\left(\frac{R}{\gamma} + 1\right).$$

The proof of Theorem 13 relies on the following lemma:

Lemma 14. *Take A as defined in (28), where each v_q is given by a different row of an (R, γ)-incoherent design. Then*

$$\mu_B(A) \le \frac{\gamma}{N}.$$

Proof. Note that for $i \neq j$, we have from (27) that

$$
\begin{aligned}
A_i^T A_j &= \sum_{k=1}^{Q} (A_{v_k})_i^T (A_{v_k})_j \\
&= \sum_{k=1}^{Q} (T^{i-1} B v_k(i) I)^T (T^{j-1} B v_k(j) I) \\
&= \left(\sum_{k=1}^{Q} v_k(i) v_k(j) \right) B^T T^{-(i-1)} T^{j-1} B = \langle V_i, V_j \rangle B^T T^{j-i} B,
\end{aligned}
$$

where V_i denotes the ith column of the (R, γ)-incoherent design matrix V. Then

$$
\begin{aligned}
\|A_i^T A_j\|_{2 \to 2} &= \left\| \langle V_i, V_j \rangle B^T T^{j-i} B \right\|_{2 \to 2} \\
&\leq |\langle V_i, V_j \rangle| \|B^T\|_{2 \to 2} \|T^{i-j}\|_{2 \to 2} \|B\|_{2 \to 2} = |\langle V_i, V_j \rangle| \leq \gamma,
\end{aligned}
$$

where the last inequality uses the fact that any pair of columns of V have at most γ ones in common. Definition 11 then gives the result:

$$
\mu_B(A) = \frac{1}{N} \max_{i \neq j} \|A_i^T A_j\|_2 \leq \frac{\gamma}{N}. \qquad \square
$$

We can now prove the main result:

Proof of Theorem 13. Note that BOMP recovers z from $w = Az$ if and only if it recovers z from $\tilde{w} = (1/\sqrt{R}) Az$. As such, it suffices to check that $\tilde{A} := (1/\sqrt{R}) A$ satisfies the conditions of Theorem 12. First, $\tilde{A}$ has columns of unit 2-norm due to the scaling. Since $\tilde{A}$ is composed of orthogonal columns, we have $\nu(\tilde{A}) = 0$. As such, it suffices to have

$$
NK < \frac{1}{2} \left(\frac{1}{\mu_B(\tilde{A})} + N \right) = \frac{1}{2} \left(\frac{R}{\mu_B(A)} + N \right).
$$

Lemma 14 then gives the result. $\qquad \square$

At this point, we note that Theorem 3 in [34] gives a construction of a $Q \times L$ $(R, 1)$-incoherent design, where R is the smallest integer satisfying $L \leq 4R^2 \log^2(2R)$ and $Q := \lceil 4R^2 \log(4R) \rceil$. As such, the number of exposures is $Q = O(L/\log R) = O(L/\log L)$, and we can have $K = O(R) = O(\sqrt{L}/\log L)$. We note that this number of exposures is a vanishing fraction of L, i.e., the number of exposures required by spatial scanning. As such, this imaging system is legitimately compressive. Also, the number of active blocks we allow scales according to a square-root bottleneck, which we should expect from a coherence-based guarantee.

4.5 Simulations

In this section, we demonstrate the utility of BOMP when applied to spectroscopy. We can break down our simulations into three steps: (1) simuating of the data cube, (2) simulating of the sensor, and (3) recovering the data.

First, we produced simulated data according to the description in Section 4.1 (see Appendix B for MATLAB code). In particular, we created a $1 \times 100 \times 100$ data cube X of stars in space. For our simulation, spectral bands range from 400 nm to 895 nm at increments of 5 nm. The temperature T of a star is chosen uniformly at random from a range of 4000K to 10000K. Each star is simulated with ten absorption lines, each at random wavelengths $\{\lambda_1, \ldots, \lambda_{10}\}$ drawn uniformly from the range of spectral bands. These absorption lines are simulated with negative Lorentzian distributions as in (29), where α_k and β_k are height and width parameters respectively for the kth absorption line. Specifically, α_k is chosen uniformly from a range of 0 to 2 nm and β_k is chosen uniformly from a range of 0 to $L(\lambda_k, T)/2$, where $L(\lambda_k, T)$ is the spectral radiance at wavelength λ_k from Definition 9. Figure 8 is an example of the star data simulated in MATLAB. Throughout the simulations, we take the block sparsity level to be $K \leq 3$, which is reasonable considering there should not be many stars in any

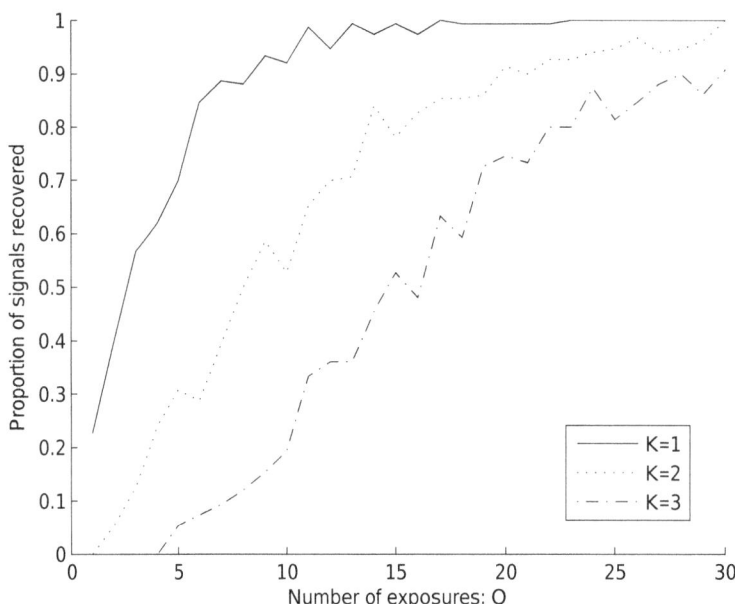

Figure 9. The proportion of 150 randomly generated $1 \times 100 \times 100$ data cubes X which are recovered using BOMP (see Appendix B for MATLAB code). The solid line shows the proportion of data cubes with block sparsity level $K = 1$ that were recovered, the dotted line shows the proportion for $K = 2$, and the dash-dotted line illustrates the $K = 3$ case.

given $1 \times 100 \times 100$ data cube slice. Once X is generated, we create a block sparse vector z from X by reshaping.

Next, we simulate taking Q exposures with the sensing platform using the measurement matrix A described in Section 4.1. We let the mirrors in the MMA have random orientation, meaning the entries of V are iid Bernoulli with some probability parameter p. As such, the columns of V will have different numbers of ones, and so V will not be an incoherent design as defined in Section 4.4. Still, we take inspiration from Theorem 13 by taking p to be somewhat small. Indeed, each column will tend to have about Qp ones, and pairs of columns will tend to have about Qp^2 ones in common. Thus, the natural proxy for R/γ is $(Qp)/(Qp^2) = 1/p$, and so Theorem 13 suggests that we should take p small in order to sense signals with large block sparsity K. For our simulations, we take $p = 1/4$.

Finally, after generating $w = Az$, we use BOMP as defined in Section 4.3 to get a solution $\tilde{z}$. In these simulations, A is $Q(N + L - 1) \times LN = 199Q \times 10000$, and we let Q range up to 30. As such, A is too large to implement naively on a standard desktop. We therefore hard-coded how each of the blocks of A (and their transposes) act on a given vector (see MATLAB code in Appendix B). This led to significant speedups in calculating $w = Az$ and in running BOMP.

Number of exposures for recovery.

In the first simulation, we generated data cubes with varying sparsity levels K and observed the performance of BOMP using $Q \in \{1, \ldots, 30\}$ exposures. For each data cube, BOMP ran for K iterations. For each value of Q, we performed BOMP on 150 random data cubes to produce the estimate $\tilde{z}$, and we declared successful recovery if $\|z - \tilde{z}\|_2 < 10^{-14}$. Figure 9 illustrates the proportion of trials that were successful for block sparsity level $K = 1$ (solid line), $K = 2$ (dotted line), and $K = 3$ (dash-dotted line). As we can see, it suffices to have $Q = 13$ exposures for successful recovery when $K = 1$. Also, 90 percent of the data cubes with block sparsity $K = 2$ and $K = 3$ are successfully recovered from 22 and 30 exposures, respectively.

The effect of noisy measurements.

In the second simulation, we add Gaussian random noise to our measurements. Specifically, we measure a data cube with block sparsity level $K = 3$ and $Q = 30$ exposures, resulting in measurements of the form $w = Az + \epsilon$, where ϵ is a vector composed of iid Gaussian random variables with mean zero and standard deviation 0.008. Figure 10(a) shows the true spectral radiance of one of the stars in z, and Figure 10(b) shows the spectral radiance of the reconstruction using K iterations of BOMP. As we can see, shallow absorption lines (located at wavelengths 560 nm, 815 nm, and 825 nm in our example) are drowned out by the noise. In practice, one

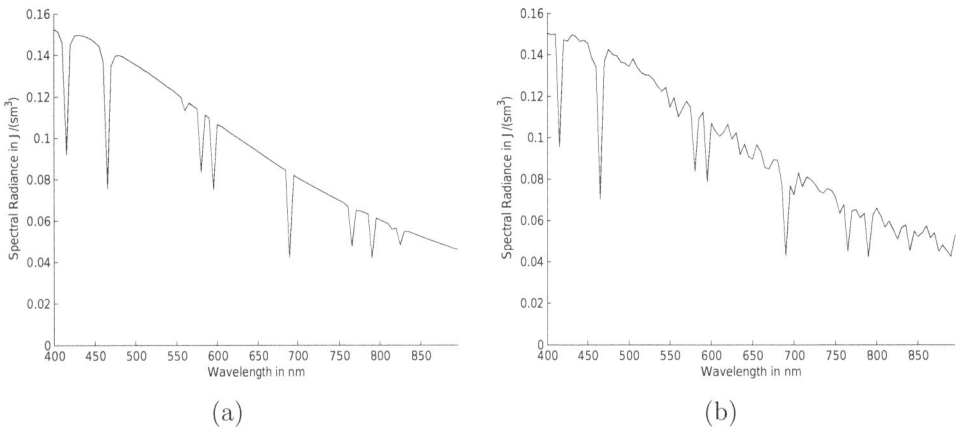

Figure 10. (a) The spectrum of one of the three simulated stars, all of which were measured in the presence of Gaussian noise. (b) Reconstruction of the same star's spectrum using BOMP. Notice that the three most shallow absorption lines (at 560, 815, and 825 nm) were lost in the noise.

might overcome this apparent loss of information by fitting spectra of chemicals to the absorption line locations and depths. Indeed, the star's chemical composition is precisely the objective of spectroscopy.

V. Conclusions and future work

In this chapter, we summarize our results and discuss future research.

5.1 Discrete uncertainty principles and applications

In Chapter II, we discussed the 0-norm uncertainty principle and gave a concise proof of the Donoho–Stark uncertainty principle developed in [8]. Additionally, we introduced and characterized equality in a new mixed-norm uncertainty principle. We also showed that a discretized and periodized Gaussian achieves near equality in both the Donoho–Stark and mixed-norm uncertainty principle. In Chapter III, we applied these results to demonstrate a fundamental limitation in signal demixing and to detect K-sparse signals from only $O(K)$ measurements.

For future work, it would be interesting to apply this new uncertainty principle to the fundamental problem of compressed sensing with a partial Fourier matrix [35, 36]. Specifically, how many random rows M of the $N \times N$ discrete Fourier transform matrix are necessary to satisfy the (K, δ)-restricted isometry property? Recently, Bourgain [36] demonstrated that $M = O_\delta(K \log^3 N)$ rows suffice, whereas Gaussian matrices only need $M = O_\delta(K \log(N/K))$ [6]. Recall that in Section 3.2, we used our uncertainty principle to show that if $\|x\|_0 \leq K$ and $\|x\|_2^2 = N/K$, then a random partial Fourier matrix A satisfies $\|Ax\|_1 \gg 0$ with high probability provided $M = O(K)$. Since the RIP problem is so similar, we suspect that similar ideas might apply.

5.2 Fast hyperspectral imaging

In Chapter IV, we discussed the hyperspectral imaging problem in the context of spectroscopy of stars. We briefly described conventional hyperspectral platforms and their limitations before proposing a compressed sensing platform that can quickly sample hyperspectral data. Using combinatorial designs, we built coded apertures that can be used to quickly collect hyperspectral data, and from which we can quickly recover the desired imagery using block orthogonal matching pursuit.

For future work in this direction, we are particularly excited about real-world implementation. This will bring its own challenges, such as calibration and non-Gaussian noise, but we suspect the theory is robust to such deviations, and it would be interesting to explicitly include them in the theory.

Also, we currently use random coded apertures, but with each exposure, the micro-mirrors need to mechanically change their orientation. In the real world, these changes in orientation cost energy, which is a limited resource in remote sensing platforms. As such, one possible area for future research would be to develop incoherent designs for which changes in orientation are minimized.

Appendix A. Supporting material

1.1 Harmonic analysis background

For this section define χ as in Section 2.1. The following claim helps establish the assumptions needed for Definition 1.

Claim 1. *The dual of G, $\widehat{G} := \{\chi\}$ is an abelian group.*

Proof. Let $g_1, g_2 \in G$ and $\chi_1, \chi_2 \in \widehat{G}$,

$$\chi_1\chi_2[g_1 + g_2] = \chi_1[g_1 + g_2]\chi_2[g_1 + g_2]$$

$$= \chi_1[g_1]\chi_1[g_2]\chi_2[g_1]\chi_2[g_2] = \chi_1[g_1]\chi_2[g_1]\chi_1[g_2]\chi_2[g_2] = \chi_1\chi_2[g_1]\chi_1\chi_2[g_2].$$

Therefore, $\chi_1\chi_2 \in \widehat{G}$ so $\widehat{G}$ is closed. Define $e(g) := 1 \ \forall g \in G$. Then

$$e[g_1 + g_2] = 1 = 1 \cdot 1 = e[g_1]e[g_2],$$

meaning $e \in \widehat{G}$, and so $\widehat{G}$ has an identity element. Define $\phi[g] := 1/\chi[g] \ \forall g \in G$

$$\phi[g_1 + g_2] = \frac{1}{\chi[g_1 + g_2]} = \frac{1}{\chi[g_1]\chi[g_2]} = \left(\frac{1}{\chi[g_1]}\right)\left(\frac{1}{\chi[g_2]}\right) = \phi[g_1]\phi[g_2].$$

Therefore, $\phi \in \widehat{G}$. Let g be an arbitrary element of G. Then

$$(\phi\chi)[g] = \phi[g]\chi[g] = \frac{1}{\chi[g]}\chi[g] = 1 = e[g],$$

and so $\phi = \chi^{-1}$. Thus, $\widehat{G}$ is a group. Lastly,

$$(\chi_1\chi_2)[g] = \chi_1[g]\chi_2[g] = \chi_2[g]\chi_1[g] = (\chi_2\chi_1)[g],$$

and so $\widehat{G}$ is an abelian group. $\qquad\qquad\qquad\qquad\qquad\qquad\qquad\qquad\qquad\quad \square$

The following Claims help establish that $\widehat{G^k} = \widehat{G}^k$.

Claim 2. *For all* $\chi \in \widehat{G}$, $\sum_{g \in G} \chi[g] = 0$.

Proof. Let $\chi \in \widehat{G}$ and fix $g_2 \in G$ such that $\chi[g_2] \neq 1$.

$$\sum_{g_1 \in G} \chi[g_1]\chi[g_2] = \sum_{g_1 \in G} \chi[g_1 + g_2] = \sum_{g_1 \in G} \chi[g_1].$$

where the last equality is true by the closure of G. Rearranging,

$$\sum_{g_1 \in G} \chi[g_1]\left(\chi[g_2] - 1\right) = 0.$$

We know $\chi[g_2] - 1 \neq 0$ because g_2 is arbitrary so $\sum_{g \in G} \chi[g] = 0$. $\square$

Because $\sum \chi[g] = 0$ we can show that $\widehat{G}$ consists of orthogonal elements of $\ell(G, \mathbb{C})$. In fact, $\widehat{G}$ is an orthonormal basis for $\ell(G, \mathbb{C})$.

Claim 3. $\{\chi\}_{\chi \in \widehat{G}}$ *is an orthonormal basis for* $\ell(G, \mathbb{C})$.

Proof. Let $\chi \in \widehat{G}$. First, we will show that $\|\chi\|_2 = 1 \; \forall \chi \in \widehat{G}$.

$$\left\langle \frac{1}{\sqrt{|G|}}\chi, \frac{1}{\sqrt{|G|}}\chi \right\rangle = \frac{1}{|G|}\sum_{g \in G} \chi[g]\overline{\chi[g]} = \frac{1}{|G|}\sum_{g \in \widehat{G}} |\chi[g]|^2 = \frac{1}{|G|}\sum_{g \in G} 1 = 1.$$

Therefore, the statement is true for an arbitrary χ so it is true for all $\chi \in \widehat{G}$. Additionally, the above shows that $\overline{\chi[g]} = \chi^{-1}[g]$. Second, we will show orthogonality. Let χ, ψ be arbitrary elements of $\widehat{G}$, such that $\chi \neq \psi$. Then,

$$\left\langle \frac{1}{\sqrt{|G|}}\chi, \frac{1}{\sqrt{|G|}}\psi \right\rangle = \frac{1}{|G|}\sum_{g \in G} \chi[g]\overline{\psi[g]} = \frac{1}{|G|}\sum_{g \in G} (\chi\psi^{-1})[g] = 0.$$

The last inequality comes from the fact that $\chi\psi^{-1} \in \widehat{G}$ and Claim 2. Therefore $\{\chi\}_{g \in \widehat{G}}$ is orthonormal. We know that $\ell(G, \mathbb{C})$ is a $|G|$-dimensional vector space

60

and that $|\widehat{G}| = |G|$. Therefore, $\{\chi\}_{\chi \in \widehat{G}}$ spans $\ell(G, \mathbb{C})$. Thus, $\{\chi\}_{\chi \in \widehat{G}}$ is a basis for $\ell(G, \mathbb{C})$. $\qquad \square$

Claim 4. *If* $\chi_1, \cdots, \chi_k \in \widehat{G}$ *then* $\chi_1 \otimes \cdots \otimes \chi_k \in \widehat{G^k}$.

Proof. By definition,

$$
\begin{aligned}
(\chi_1 \otimes \cdots \otimes \chi_k)[(g_1, g_2, \cdots, g_k) + (g_1', g_2', \cdots, g_k')] &= (\chi_1 \otimes \cdots \otimes \chi_k)[g_1 + g_1', \cdots, g_k + g_k'] \\
&= \chi_1[g_1 + g_1'] \cdots \chi_k[g_k + g_k'] \\
&= \chi_1[g_1]\chi_1[g_1'] \cdots \chi_k[g_k]\chi_k[g_k'] \\
&= (\chi_1[g_1] \cdots \chi_k[g_k])(\chi_1[g_1'] \cdots \chi_k[g_k']).
\end{aligned}
$$

The last equality is the definition of $(\chi_1 \otimes \cdots \otimes \chi_k)[g_1, \cdots, g_k](\chi_1 \otimes \cdots \otimes \chi_k)[g_1' \cdots g_k']$. Therefore, $\chi_1 \otimes \cdots \otimes \chi_k \in \widehat{G^k}$. $\qquad \square$

We have thus proven that $\{\chi_1 \otimes \chi_2 \otimes \cdots \otimes \chi_k\} \subset \widehat{G^k}$ but we can actually prove set equality.

Claim 5. *For* $\chi_1, \chi_2, \cdots, \chi_k \in \widehat{G}$, $\{\chi_1 \otimes \chi_2 \otimes \cdots \otimes \chi_k\} = \widehat{G^k}$

Proof. We need to show that $\#\{\chi_1 \otimes \chi_2 \otimes \cdots \otimes \chi_k\} = |\widehat{G^k}|$. Hence, it suffices to show that each $\chi_1 \otimes \chi_2 \otimes \cdots \otimes \chi_k$ is unique. Assume $\chi_1 \otimes \chi_2 \otimes \cdots \otimes \chi_k[h_1, h_2, \cdots h_k] = \psi_1 \otimes \psi_2 \otimes \cdots \otimes \psi_k[h_1, h_2, \cdots, h_k]$ for all $[h_1, h_2, \cdots, h_k] \in G^k$. Let $[g_1, g_2, \cdots, g_k] \in G^k$ such that $g_m = g \in G$ and $g_n = 0$ for all $n \neq m$. Therefore, by definition

$$(\chi_1 \otimes \chi_2 \otimes \cdots \otimes \chi_k)[g_1, g_2, \cdots, g_k] = (\psi_1 \otimes \psi_2 \otimes \cdots \otimes \psi_k)[g_1, g_2, \cdots, g_k]$$

$$\chi_1[g_1]\chi_2[g_2] \cdots \chi_k[g_k] = \psi_1[g_1]\psi_2[g_2] \cdots \psi_k[g_k]$$

Plugging in our definition of $(g_1, g_2, \cdots, g_k)$ we get

$$\chi_1[0]\chi_2[0]\cdots\chi_m[g]\cdots\chi_k[0] = \psi_1[0]\psi_2[0]\cdots\psi_m[g]\cdots\psi_k[0]$$

$$\chi_m[g] = \psi_m[g].$$

Since g and m are arbitrary, this statement is true for all $g \in G$ and $m \leq k$. Thus, $\chi_1 \otimes \chi_2 \otimes \cdots \otimes \chi_k = \psi_1 \otimes \psi_2 \otimes \cdots \otimes \psi_k$. So each $\chi_1\otimes, \chi_2 \otimes \cdots \otimes \chi_k$ is unique. $\qquad \square$

By Claims 3 and 5 we see that $\widehat{G^k} = \widehat{G}^k$.

1.2   Poisson summation formula

The use of the Poisson Summation Formula relies on specific assumptions regarding f. It suffices that the function f is in Schwarz Space. A consiquence of $f \in \mathcal{S}$ is that $f(x) \leq C/x^2$ for some $C > 0$. A quick review of the bounded convergence theorem will be useful as we build the function that achieves near equality in Theorems 2 and 6. The theorem below is a corollary to the bounded convergence theorem.

Lemma 15. *Assume,*

$$f(x) = \sum_{n=-\infty}^{\infty} h(x_n), \qquad f_N(x) := \sum_{n=-N}^{N} h(x+n)$$

and f_N converges pointwise to f. Then

$$\lim_{N\to\infty} \int_0^1 f_N(x)dx = \int_0^1 f(x)dx.$$

Proof. By the triangle inequality and Definition 2, we know that

$$|f_N(y)| = \left| \sum_{n=-N}^{N} h(x+n) \right| \leq \sum_{n=-N}^{N} |h(x+)| \leq \sum_{n=-N}^{N} \frac{C}{(x+n)^2+1} = C \sum_{n=-N}^{N} \frac{1}{(x+n)^2+1}.$$

We can break the sum up into negative and nonnegative indices. Hence

$$C \sum_{n=-N}^{N} \frac{1}{(x+n)^2 + 1} = C \left[\sum_{n=-N}^{-1} \frac{1}{(x+n)^2 + 1} + \sum_{n=0}^{N} \frac{1}{(x+n)^2 + 1} \right].$$

Since $x \in [0, 1]$,

$$C \left[\sum_{n=-N}^{-1} \frac{1}{(x+n)^2 + 1} + \sum_{n=0}^{N} \frac{1}{(x+n)^2 + 1} \right] \leq C \left[\sum_{n=-N}^{-1} \frac{1}{(n+1)^2 + 1} + \sum_{n=0}^{N} \frac{1}{n^2 + 1} \right].$$

Let $\tilde{n} := -n$. Then

$$C \left[\sum_{n=-N}^{-1} \frac{1}{(n+1)^2 + 1} + \sum_{n=0}^{N} \frac{1}{n^2 + 1} \right] = C \left[\sum_{\tilde{n}=1}^{N} \frac{1}{(\tilde{n}-1)^2 + 1} + \sum_{n=0}^{N} \frac{1}{n^2 + 1} \right].$$

Combining terms gives us

$$C \left[\sum_{\tilde{n}=1}^{N} \frac{1}{(\tilde{n}-1)^2 + 1} + \sum_{n=0}^{N} \frac{1}{n^2 + 1} \right] = C \left[1 + \sum_{n=1}^{N} \left(\frac{1}{(n-1)^2 + 1} + \frac{1}{n^2 + 1} \right) \right].$$

Since $n > 0$,

$$C \left[1 + \sum_{n=1}^{N} \left(\frac{1}{(n-1)^2 + 1} + \frac{1}{n^2 + 1} \right) \right] \leq C \left[1 + 2 \sum_{n=1}^{N} \frac{1}{(n-1)^2 + 1} \right].$$

Let $n' = n - 1$. Then by change of variables,

$$C \left[1 + 2 \sum_{n'=0}^{N-1} \frac{1}{(n')^2 + 1} \right] \leq C \left[1 + 2 + 2 \int_0^{N-1} \frac{1}{y^2 + 1} dy \right] \leq C \left[3 + 2 \int_0^{\infty} \frac{1}{y^2 + 1} dy \right].$$

Evaluating the integral then gives

$$C \left[3 + 2 \int_0^{\infty} \frac{1}{y^2 + 1} dy \right] = C \left[3 + 2 \arctan(y) \Big|_0^{\infty} \right] = C \left[3 + \pi \right] =: M.$$

Therefore, $\exists M \in \mathbb{R}$ such that $|f_N(y)| \leq M$. Thus, by the bounded convergence theorem,

$$\lim_{N \to \infty} \int_0^1 f_n(x)dx = \int_0^1 f(x)dx \qquad \square$$

Another tool necessary to prove the Poisson summation formula is Lemma 16 below.

Lemma 16. *If $f \in \mathcal{S}$, then*

$$\int_0^1 \sum_{n \in \mathbb{Z}} f(x+n) e^{-2\pi i k x} dx = \sum_{n \in \mathbb{Z}} \int_0^1 f(x+n) e^{-2\pi i k x} dx.$$

Proof. Define $h_N(x) := f_N(x)e^{-2\pi i k x}$ where $f_N(x)$ is defined as in Lemma 15. We know that $h_N(x)$ converges pointwise to $f(x)e^{-2\pi i k x}$. Thus

$$|h_N(x)| = |f_N(x)e^{-2\pi i k x}| = |f_N(x)||e^{-2\pi i k x}| \leq M.$$

Therefore, by Lemma 15

$$\lim_{N \to \infty} \int_0^1 \sum_{n=-N}^{N} g(x+n) e^{-2\pi i k x} dx = \int_0^1 \sum_{n \in \mathbb{Z}} g(x+n) e^{2\pi i k x} dx. \qquad (33)$$

Additionally, we know

$$\int_0^1 h_N(x)dx = \int_0^1 \sum_{n=-N}^{N} g(x+n) e^{-2\pi i k x} dx = \sum_{n=-N}^{N} \int_0^1 g(x+n) e^{-2\pi i k x} dx. \qquad (34)$$

Taking the limit of both sides of (34) we get

$$\lim_{N \to \infty} \int_0^1 \sum_{n=-N}^{N} g(x+n) e^{-2\pi i k x} dx = \sum_{n \in \mathbb{Z}} \int_0^1 g(x+n) e^{-2\pi i k x} dx \qquad (35)$$

Therefore, by (33) and (35) yields,

$$= \sum_{n \in \mathbb{Z}} \int_0^1 g(x+n) e^{-2\pi i k x} dx = \int_0^1 \sum_{n \in \mathbb{Z}} g(x+n) e^{2\pi i k x} dx. \qquad \square$$

With Lemma 16 we have a sufficient foundation to prove the Poisson summation formula [37].

Theorem 14 (Poisson Summation Formula). *For some* $g \in \mathcal{S}$,

$$\sum_{n \in \mathbb{Z}} g[n] = \sum_{k \in \mathbb{Z}} (F_{\mathbb{R}} g)[k]$$

Proof. Define $f(x) := \sum_{n \in \mathbb{Z}} g(x+n)$ and $F_{\mathbb{T}}$ as the Fourier transform from $\mathbb{R}$ to $\mathbb{Z}$. Then,

$$\sum_{n \in \mathbb{Z}} g(x+n) = (F_{\mathbb{T}}^{-1} F_{\mathbb{T}} f)(x) = \sum_{k \in \mathbb{Z}} (F_{\mathbb{T}} f)[k] e^{2\pi i k x} = \sum_{k \in \mathbb{Z}} \frac{1}{\sqrt{2\pi}} \int_0^{2\pi} f(x) e^{-2\pi i k x} dx e^{2\pi i k x}.$$

Substituting in our definition of $f(x)$ yields

$$\sum_{k \in \mathbb{Z}} \frac{1}{\sqrt{2\pi}} \int_0^{2\pi} f(x) e^{-2\pi i k x} dx e^{2\pi i k x} = \sum_{k \in \mathbb{Z}} \frac{1}{\sqrt{2\pi}} \int_0^{2\pi} \left[\sum_{n \in \mathbb{Z}} g(x+n) \right] e^{-2\pi i k x} dx e^{2\pi i k x}$$

$$\tag{36}$$

$$= \sum_{k \in \mathbb{Z}} \frac{1}{\sqrt{2\pi}} \sum_{n \in \mathbb{Z}} \int_0^{2\pi} g(x+n) e^{-2\pi i k x} dx e^{2\pi i k x}.$$

Where the last statement is true by Lemma 16. Substituting $u = n + x$ into (36) gives us

$$\sum_{k \in \mathbb{Z}} \frac{1}{\sqrt{2\pi}} \sum_{n \in \mathbb{Z}} \int_n^{2\pi + n} g(u) e^{-2\pi i k(u-n)} du e^{2\pi i k x}. \tag{37}$$

65

We know that

$$\sum_{n\in\mathbb{Z}}\int_n^{2\pi+n} g(u)e^{-2\pi ik(u-n)}du = \sum_{n\in\mathbb{Z}}\int_n^{2\pi+n} g(u)e^{-2\pi ik}e^{2\pi ikn}du = \int_{-\infty}^{\infty} g(u)e^{-2\pi iku}du.$$

because $e^{2\pi in} = 0$. Thus, (37) is equivalent to

$$\sum_{k\in\mathbb{Z}}\left[\frac{1}{\sqrt{2\pi}}\int_{-\infty}^{\infty} g(u)e^{-2\pi iku}du\right]e^{2\pi ikx} = \sum_{k\in\mathbb{Z}}(F_\mathbb{R}g)[k]e^{2\pi ikx}$$

where the last equality is by the definition of $F_\mathbb{R}g$. Therefore,

$$\sum_{n\in\mathbb{Z}} g(x+n) = \sum_{k\in\mathbb{Z}}(F_\mathbb{R}g)[k]e^{2\pi ikx}.$$

Evaluating both sides at $x = 0$, we get

$$\sum_{n\in\mathbb{Z}} g[n] = \sum_{k\in\mathbb{Z}}(F_\mathbb{R}g)[k]. \qquad \square$$

Theorem 14 will be useful for finding the Fourier transform pair needed for near equality.

Corollary 1. *For all $f \in \mathcal{S}$,*

$$\sum_{p\in\mathbb{Z}} f(p-s) = \sum_{q\in\mathbb{Z}}(F_\mathbb{R}f)(q)e^{-2\pi iqs}$$

Proof. Let, $h(x) := f(x-s)$. By definition,

$$(F_\mathbb{R}h)(x) := \int_{-\infty}^{\infty} h(t)e^{-2\pi ixt}dt. \tag{38}$$

Substituting $t' = t - s$ into (38) yields

$$\int_{-\infty}^{\infty} h(t)e^{-2\pi ixt}dt = \int_{-\infty}^{\infty} f(t-s)e^{-2\pi ixt}dt = \int_{-\infty}^{\infty} f(t')e^{-2\pi ix(t'+s)}dt'.$$

We know that $y^{a+b} = y^a y^b$. Therefore

$$\int_{-\infty}^{\infty} f(t')e^{-2\pi ix(t'+s)}dt' = e^{-2\pi ixs}\int_{-\infty}^{\infty} f(t')e^{-2\pi ixt'}dt' = e^{-2\pi ixs}\left(F_{\mathbb{R}}f\right)(x). \qquad (39)$$

By Theorem 14,

$$\sum_{p\in\mathbb{Z}} h(p) = \sum_{q\in\mathbb{Z}}\left(F_{\mathbb{R}}h\right)(q).$$

Substituting in our definition of h and the result of (39),

$$\sum_{p\in\mathbb{Z}} f(p-s) = \sum_{q\in\mathbb{Z}}\left(F_{\mathbb{R}}f\right)(q)e^{-2\pi qs} \qquad\qquad \square$$

1.3   Properties of the Gaussian

Claim 6. *for all* $y \in \mathbb{C}$

$$\int_{-\infty}^{\infty} e^{-y^2}dy = \sqrt{\pi} \qquad (40)$$

Proof. Because $x = \sqrt{x^2}$, we can rewrite (40) as

$$\int_{-\infty}^{\infty} e^{-y^2}dy = \left(\int_{-\infty}^{\infty} e^{-y^2}dy \int_{-\infty}^{\infty} e^{-x^2}dx\right)^{1/2} = \left(\int_{-\infty}^{\infty}\int_{-\infty}^{\infty} e^{-(y^2+x^2)}dydx\right)^{1/2}.$$

Converting to polar coordinates yields

$$\left(\int_{-\infty}^{\infty}\int_{-\infty}^{\infty} e^{-(y^2+x^2)}dydx\right)^{1/2} = \left(\int_{0}^{2\pi}\int_{0}^{\infty} e^{-r^2}rdrd\theta\right)^{1/2}.$$

Let $u := r^2$. Then by change of variables

$$\left(\int_0^{2\pi} \int_0^\infty e^{-r^2} r\, dr\, d\theta \right)^{1/2} = \left(\frac{1}{2} \int_0^{2\pi} \int_0^\infty e^{-u}\, du\, d\theta \right)^{1/2} = \left(\pi \int_0^\infty e^{-u}\, du \right)^{1/2} = \sqrt{\pi}$$

which implies the result. $\square$

Given the integral of e^{-y^2} we can easily compute $F(e^{-y^2})$.

Claim 7. *The Fourier transform of $f(x) := e^{-x^2}$ is*

$$(Ff)(\xi) = e^{-\pi^2 \xi^2} \sqrt{\pi}$$

Proof. By definition of Fourier transform,

$$(Ff)(\xi) = \int_{-\infty}^\infty e^{-x^2} e^{-2\pi i \xi x}\, dx = \int_{-\infty}^\infty e^{-(x+2\pi i \xi x)}\, dx = \int_{-\infty}^\infty e^{-(x+\pi i \xi)^2 - \pi^2 \xi^2}\, dx$$

where the last equality is from completing the square. Let $u := -x\pi i \xi$.

Since f is analytic,

$$\int_{-\infty}^\infty e^{-(x+\pi i \xi)^2 - \pi^2 \xi^2}\, dx = e^{-\pi^2 \xi^2} \int_{-\infty}^\infty e^{-u^2}\, du = \sqrt{\pi} e^{-\pi^2 \xi^2}$$

where the last inequality is due to Claim 6. $\square$

One specific property of the Fourier transform pair defined in Claim 7 is the property of dilation.

Claim 8. *If $h(x) := f(mx)$ for some function f, then*

$$(Fh)(\xi) = \frac{(Ff)\left(\frac{\xi}{m}\right)}{m}$$

Proof. By Definition of $F : \mathbb{C} \to \mathbb{C}$ and h,

$$(Fh)(\xi) = \int_{-\infty}^{\infty} h(x)e^{-2\pi i \chi x} dx = \int_{-\infty}^{\infty} f(mx)e^{-2\pi i \chi x} dx = \frac{1}{m} \int_{-\infty}^{\infty} f(u)e^{-2\pi i \xi \frac{u}{m}} du$$

where the last equality comes from change of variables where $u := x/m$. Rearranging gives us the result:

$$\frac{1}{m} \int_{-\infty}^{\infty} f(u)e^{-2\pi i u \frac{\xi}{m}} du = \frac{1}{m}(Ff)\left(\frac{\xi}{m}\right). \qquad \square$$

1.4 Supporting claim for Section 2.4

Claim 9 (Geometric Sum Formula). *Suppose $q, k \in \mathbb{Z}$. Then*

$$\sum_{j=0}^{n-1} \left[e^{2\pi i(q-k)/n}\right]^j = \begin{cases} n, & \frac{q-k}{n} \in \mathbb{Z}, \\ 0, & \text{otherwise}. \end{cases}$$

Proof. Without loss of generality, assume that $e^{2\pi i(q-k)/n} \neq 1$. Then

$$\sum_{j=0}^{n-1} \left[e^{2\pi i(q-k)/n}\right]^j = \frac{\left[e^{2\pi i(q-k)/n}\right]^n - 1}{e^{\frac{2\pi i(q-k)}{n}} - 1} = \frac{e^{2\pi i(q-k)} - 1}{e^{\frac{2\pi i(q-k)}{n}} - 1} = \frac{1 - 1}{e^{\frac{2\pi i(q-k)}{n}} - 1} = 0.$$

If $e^{2\pi i(q-k)/n} = 1$, then

$$\sum_{j=0}^{n-1} \left[e^{2\pi i(q-k)/n}\right]^j = \sum_{j=0}^{n-1} 1 = n.$$

We know that $e^{2\pi i(q-k)/n} = 1$ if and only if $\frac{1-k}{n} \in \mathbb{Z}$. Therefore,

$$\sum_{j=0}^{n-1} \left[e^{2\pi i(q-k)/n}\right]^j = \begin{cases} n, & \frac{q-k}{n} \in \mathbb{Z}, \\ 0, & \text{otherwise}. \end{cases}$$

completing the proof. $\qquad \square$

Appendix B. MATLAB code for Chapter IV

Data simulation

The following MATLAB code simulates the data as described in Section 4.5.

```
function x=stardata()
% author: 2d Lt Megan Lewis
% description: outputs the data that simulates a "slice" of the sky
%  with K stars
M=100; % number of locations
L=100; % number of wavelengths
Fslice=zeros(M,L);
permindices=randperm(M);
K=2; % number of stars (K-sparse)
list=permindices(1:K); % locations of stars



% first we build a slice of the data cube
for kk=1:K
    h=6.6260*10^(-34); % planck's constant
    c=2.9979*10^8; % meters per second
    k=1.3806*10^(-23); % boltzmann's constant
    T=rand*6000+4000; % Kelvin
    lambda=(400:5:895)*10^(-9); % meters
    stuff=h*c./(lambda*k*T);
    radiance=2*h*c^2./(lambda.^5.*(exp(stuff)-1));
    radabs=radiance;
```

```
    lines=10; %number of absorbtion lines

    for ell=1:lines

        index=ceil(rand*100);

        lambda0=lambda(index);

        width=rand*2*10^(-9);

        height=rand*radiance(index)/2;

        radabs=radabs-height./(1+(lambda-lambda0).^2/width^2);

    end %for loop

    radabs=radabs/norm(radabs);

    Fslice(list(kk),:)=radabs;

end %for loop



x=Fslice';



end %function
```

Runme file

The following MATLAB code was used in order to perform block orthogonal matching pursuit (BOMP) for simulated data z and to evaluate if BOMP successfully reconstructed z.

```
%------------------------------------------------------------------------
% author: 2d Lt Megan Lewis
% description: The purpose of this .m file is to run all the functions
% necessary to generate test data, perform BOMP and evaluate the
% performance of the algorithm.
```

```matlab
%Initialization

n=0;

%loop that controls the number of masks or measurements.

for numasks=1:30

    %loop that controls the number of iterations per given number of masks.

    for runs=1:30

        n=n+1;

        %Simulates the data that BOMP is trying to recover

        z=stardata(3);


        [nz, zs]=efficient_bomp(z,numasks, 3);

        stars3(n)=norm(zs-nz);

    end %run count

end %num masks loop
```

BOMP

The following MATLAB code runs BOMP by hard-coding the model matrix.

```matlab
function [nz, zs]=efficient_bomp(z, Q, star)
%------------------------------------------------------------------------
% Author: 2d Lt Megan Lewis

% Purpose: To do Block orthogonal matching pursuit in a timely manner.

%------------------------------------------------------------------------
% INPUT

% z=vectorized data cube

% k=sparsity level of z

% Q=number of exposures
```

```
%-------------------------------------------------------------------------
% OUTPUT
% nz=recovered z
%-------------------------------------------------------------------------
% INITIALIZATION
[L, N]=size(z);
zs=reshape(z,[numel(z),1]); %turns data cube into data vector
w=zeros(Q*(N+L-1),1); %initiates w
%The incoherent design we will be taking our mask values from:
V=binornd(1,0.25,Q,L);
%-------------------------------------------------------------------------
% Getting w=Az
for q=1:Q
    for l=1:L
        w((q-1)*(N+L-1)+1:q*(N+L-1))=w((q-1)*(N+L-1)+1:q*(N+L-1))...
            +V(q,l)*[zeros(l-1,1); zs((l-1)*N+1:l*N); zeros(L-l,1)];
    end %l for loop
end %q for loop
%-------------------------------------------------------------------------
%INITIALIZATION OF BOMP
bestinds=[];
Id=eye(N);
r=w;
%-------------------------------------------------------------------------
% BOMP
for index=1:star
```

```
%First find argmax||A_i^*r||_2
norms=zeros(L,1);
for l=1:L
    temp=0;
    for q=1:Q
        temp=temp+V(q,l)*r(1+(N+L-1)*(q-1):l+N-1+(N+L-1)*(q-1));
    end %q for loop
    norms(l)=norm(temp);



end %l for loop
[val ind]=max(norms);
bestinds(index)=ind;
%Next we calculate Aind
Aind=zeros(Q*(N+L-1),length(bestinds)*N);
for ii=1:length(bestinds)
    jj=bestinds(ii);
    for n=1:N
        for q=1:Q


            Aind((q-1)*(N+L-1)+1:q*(N+L-1),(ii-1)*N+n)=...
                Aind((q-1)*(N+L-1)+1:q*(N+L-1),(ii-1)*N+n)+V(q,jj)*...
                [zeros(jj-1,1); Id(:,n); zeros(L-jj,1)];
        end %q for loop
```

```
        end %n for loop

    end %ii for loop

    %Calculating ztilde

    ztilde=Aind\w; %least squares minimization

    % Updating the residuals

    r=w-Aind*ztilde;

end % index for loop

%-----------------------------------------------------------------------

% Calculating nz

nz=zeros(size(zs));

for ii=1:length(bestinds)

    kk=bestinds(ii);

    nz((kk-1)*N+1:kk*N)=ztilde((ii-1)*N+1:ii*N);

end %for loop


end %efficient_bomp
```

Bibliography

1. E. J. Candes, J. Romberg, and T. Tao, "Robust uncertainty principles: Exact reconstruction from highly incomplete frequency information," *IEEE Transactions on Information Theory*, vol. 52, pp. 489–509, 2006.

2. J. Ellenberg, "Fill in the blanks: Using math to turn lo-res datasets into hi-res samples," *Wired*, February 2010.

3. "Jpeg2000." `http://www.jpeg.org/jpeg2000`. Accessed: 2015-02-13.

4. E. J. Candes and T. Tao, "Near optimal signal recovery from random projections: universal encoding strategies?," *IEEE Transactions on Information Theory*, vol. 52, no. 12, pp. 5406–5425, 2006.

5. E. J. Candes, "The restricted isometry property and its implications for compressed sensing," *C. R. Acad. Sci. Paris, Ser. I*, vol. 346, no. 9, pp. 589–592, 2008.

6. R. Baraniuk, M. Davenport, R. DeVore, and M. Wakin, "A simple proof of the restricted isometry property for random matrices," *Constructive Approximation*, vol. 28, no. 3, pp. 253–263, 2008.

7. G. B. Folland and A. Sitaram, "The uncertainty principle: a mathematical survey," *Journal of Fourier Analysis and Applications*, vol. 3, no. 3, 1997.

8. D. L. Donoho and P. B. Stark, "Uncertainty principles and signal recovery," *SIAM Journal on Applied Mathematics*, vol. 49, no. 3, pp. 906–931, 1989.

9. S. Foucart and H. Rauhut, *A Mathematical Introduction to Compressive Sensing*. Applied and Numerical Harmonic Analysis, Springer, 2013.

10. J. A. Tropp, "On the linear independence of spikes and sines," *Journal of Fourier Analysis and Applications*, vol. 14, pp. 838–858, 2008.

11. D. J. Cooke, "A discrete X-ray transform for chromotomographic hyperspectral imaging," Master's thesis, Air Force Institute of Technology, March 2013.

12. J. A. Orson, W. F. Bagby, and G. P. Perram, "Infrared signatures from bomb detonations," *Infrared Physics and Technology*, vol. 44, pp. 101–107, 2003.

13. Q. Li, X. He, Y. Wang, H. Liu, D. Xu, and F. Guo, "Review of spectral imaging technology in biomedical engineering: achievements and challenges," *Journal of Biomedical Optics*, vol. 18, no. 10, 2013.

14. W. Cooke, "Comet Hartley 2 approaches earth," *NASA*.

15. M. F. Duarte, M. A. Davenport, D. Takhar, J. N. Laska, T. Sun, K. E. Kelly, and R. G. Baraniuk, "Single-pixel imaging via compressive sampling," *IEEE Signal Processing Magazine*, vol. 25, no. 2, p. 83, 2008.

16. Y. C. Eldar, P. Kuppinger, and H. Bolcskei, "Block-sparse signals: Uncertainty relations and efficient recovery," *IEEE Transactions on Signal Processing*, vol. 58, pp. 3042–3054, June 2010.

17. Y. C. Eldar and M. Mishali, "Robust recovery of signals from a structured union of subspaces," *IEEE Transactions on Information Theory*, vol. 55, no. 11, 2009.

18. M. B. McCoy and J. A. Tropp, "Sharp recovery bounds for convex deconvolution, with applications," *CoRR*, vol. abs/1205.1580, 2012.

19. T. Tao, "An uncertainty principle for cyclic groups of prime order," *Math Research Letters*, vol. 12, pp. 121–127, 2005.

20. M. E. Lopes, "Estimating unknown sparsity in compressed sensing," *CoRR*, vol. abs/1204.4227, 2012.

21. T. Tao, "Trick wiki article: The tensor power trick," *What's New*, 2008.

22. D. Donoho and M. Elad, "Optimally sparse representation in general (nonorthogonal) dictionaries via ℓ_1 minimization," *Proceedings of the National Academy of Sciences of the United States of America*, vol. 100, no. 5, pp. 2197–2202, 2003.

23. J. Cahill and D. G. Mixon, "Robust width: A characterization of uniformly stable and robust compressed sensing," *CoRR*, vol. abs/1408.4409, 2014.

24. B. S. Kashin and V. N. Temlyakov, "A remark on compressed sensing," *Math. Notes 82*, pp. 748–755, 2007.

25. N. Geiling, "The women who mapped the universe and still couldn't get any respect," *Smithsonian Magazine*, September 2013.

26. "The Sloan digital sky survey." `http://skyserver.sdss.org/`. Accessed: 2015-02-15.

27. N. Hagen and M. W. Kudenov, "Review of snapshot spectral imaging technologies," *Optical Engineering*, vol. 52, pp. 090901–090901, 2013.

28. M. E. Gehm, R. John, D. J. Brady, R. M. Willett, and T. J. Schulz, "Single-shot compressive spectral imaging with a dual-disperser architecture," *Applied Optics*, vol. 15, no. 21, pp. 14013–14027, 2007.

29. A. Wagadariker, R. John, R. Willett, and D. Brady, "Single disperser design for coded aperture snapshot spectral imaging," *Applied Optics*, vol. 47, no. 10, pp. B44–B51, 2008.

30. J. Caniou, *Passive Infrared Detection: Theory and Applications*. 1999.

31. J. C. D. Brand, *Lines of Light: The Sources of Dispersive Spectroscopy, 1800-1930*. Routledge, 1995.

32. C. Ott, A. Kaldun, P. Raith, K. Meyer, M. Laux, J. Evers, C. H. Keitel, C. H. Greene, and T. Pfeifer, "Lorentz meets fano in spectral line shapes: A universal phase and its laser control," *Science*, vol. 340, no. 6133, pp. 716–720, 2013.

33. J. Pitha and R. N. Jones, "Comparison of optimization methods for fitting curves to infrared band envelopes," *Canadian Journal of Chemistry*, vol. 44, pp. 3031–3050, 1966.

34. C. Beideman and J. Blocki, "Set families with low pairwise intersection," *CoRR*, vol. abs/1404.4622, 2014.

35. M. Rudelson and R. Vershynin, "On sparse reconstruction from fourier and gaussian measurements," *Communications on Pure and Applied Mathematics*, vol. 61, pp. 1025–1045, August 2008.

36. J. Bourgain, "An improved estimate in the restricted isometry problem," *Geometric Aspects of Functional Analysis*, vol. 2116, pp. 65–70, 2014.

37. J. J. Benedetto and G. Zimmermann, "Sampling multipliers and the poisson summation formula," *Journal of Fourier Analysis and Applications*, vol. 5, no. 3, pp. 505–523, 1997.